D1159052

INVITATION TO NUMBER THEORY WITH PASCAL

OTHER BOOKS OF INTEREST BY THE SAME AUTHOR

EXPLORING NUMBER THEORY WITH MICROCOMPUTERS

COMPUTER MATHEMATICS WITH BASIC PROGRAMMING

COMPUTER MATHEMATICS WITH PASCAL PROGRAMMING

NUMBER PROBLEMS AND COMPUTERS

CHALLENGING MATHEMATICAL PROBLEMS WITH BASIC SOLUTIONS

CHALLENGING MATHEMATICAL PROBLEMS WITH PASCAL SOLUTIONS

FUN WITH MICROCOMPUTERS AND BASIC

INVITATION TO NUMBER THEORY WITH PASCAL

DONALD D. SPENCER
Educational Consultant

CAMELOT PUBLISHING COMPANY
Ormond Beach, Florida

DEDICATION

To my brother, Bob

Library of Congress Cataloging-in-Publication Data

Spencer, Donald D.
Invitation to number theory with Pascal.

Includes index.

1. Numbers, Theory of--Data processing. 2. Pascal (Computer program language) I. Title.
QA241.S634 1989 512'.7 88-35344
ISBN 0-89218-126-5

Printed and bound in the United States of America

CAMELOT PUBLISHING COMPANY
P.O. Box 1357
Ormond Beach, FL 32075

PREFACE

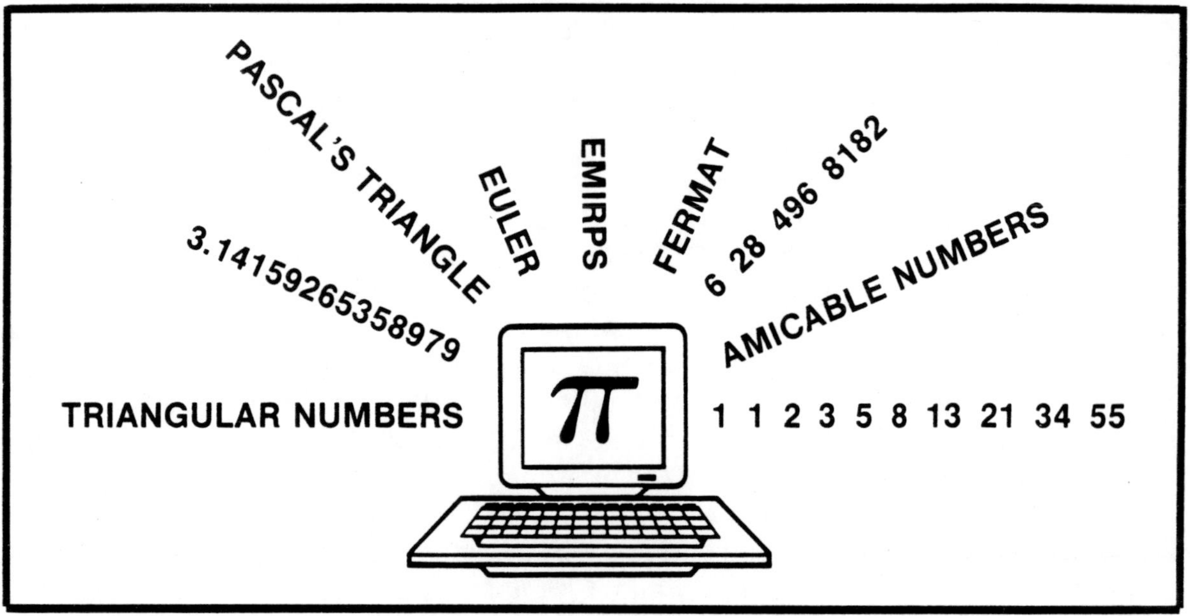

The microcomputer is one of the most important calculating tools available to mathematicians. It can take much of the drudgery out of many mathematical analyses. Many problems in mathematics require the speed and accuracy of a fast calculating device. The microcomputer is designed to handle such problems quickly and accurately.

Many complex number theory problems, as well as those of a more routine nature, can be solved conveniently with a microcomputer. Moreover, the search for solutions to research problems is greatly aided by computers. Many methods of calculation that are complex or tedious when done by hand can be accomplished with ease on a microcomputer.

Numbers have fascinated both professional mathematicians and technical people for thousands of years. Number theory concerns the most familiar of all numbers, the numbers we count with, and it may surprise you to realize how much you already know about them from experience. It is perhaps the only branch of mathematics where there is any possibility that new and valuable discoveries might be made without an extensive acquaintance with technical mathematics. In recent years, high school students have made new and important discoveries with perfect numbers and prime numbers. College students are regularly making important contributions to the number theory field.

Some reasons to use microcomputers in number theory are:

- To provide experimental evidence from which you can discover facts and theorems and make further conjectures.
- To provide reinforcement and deeper understanding of theorems through writings of programs which require understanding of proofs.
- To provide illustrations of theorems.
- To eliminate the tedious calculations that are often required.
- To provide different settings in which readers can apply number theory concepts and techniques.

The purpose of this book is to introduce the reader to elementary number theory and to some of the computer applications that the subject suggests. It can serve as a text for a beginning number theory or computer mathematics course for students with a good background in high school algebra, as a computer supplement to a more advanced treatment of number theory, or as a text for a mathematically oriented computer programming course. It can also be used as a general interest book for anyone interested in computerized number theory. The development of programs for number theory problems involves using a wide range of programming techniques, and I consider that teachers of computing could find this text very useful, providing an interesting background to many essential methods. It could well be used for a course in programming running in parallel with a course in number theory or advanced mathematics.

This text does not survey the whole area of number theory, but rather introduces many individual topics, some of which the reader may want to go into more thoroughly later. Some of the topics included are prime numbers, Fibonacci numbers, perfect numbers, palindromes, factorization, π, magic squares, modular arithmetic, Pythagorean triples, Pascal's triangle, hundreds of interesting number relationships, and recreational mathematics.

No prior experience with computers is assumed. Chapter 1 introduces the reader to computer equipment while Chapter 12 includes the essentials of the Pascal programming language. Chapters 2 through 11 present basic numerical methods and the application of these methods to number theory problems that can be conveniently programmed.

Actually, many of the methods presented can be used for calculation even when a microcomputer is not available. The last chapter in the book is designed to provide readers with a working knowledge of the Pascal language so that they will be able to understand the many programs presented in the book and write their own programs. Readers that are familiar with solving mathematical or scientific problems using the Pascal language may wish to skip this chapter.

All chapters are, for the most part, independent of one another. Objectives are listed at the beginning of each chapter. Review exercises are included at the end of each chapter. Key terms are in boldface throughout the text. Wherever possible, equipment, concepts and famous mathematicians are illustrated by drawings or computer images. To promote greater reader interest, number theory trivia items are featured in the page margin throughout the book. These features cover a wide range of topics that readers will find both educational and appealing. Included at the back of the book are a glossary of number theory and computer terms, a list of Pascal programs, a chronology of important number theory and computer events, answers to selected exercises, and an index.

The Pascal programming language is very popular and is available on most microcomputers, minicomputers and mainframe computers. All the programs in this book are written in Pascal. Since Pascal is easy to learn and use, it serves as a good introduction to computer communication. The reader is encouraged to write programs to solve problems ranging from calculating Pythagorean triples to generating magic squares. In this manner, the reader can learn the programming and numerical techniques by application to number theory problems. The book contains 95 Pascal programs to solve number theory problems. These sample computer programs are an essential feature of the book. The word "sample" should be emphasized since there are usually many programs that will perform the same task. These sample programs have been written more toward straightforward presentation, ease of comprehension, and instructional value, than toward efficiency. There was no attempt to introduce those sophisticated mathematical and programmatical techniques by which large computers can be set to the formidable task of discovering new results in number theory.

All the programs in this book were executed on IBM PC, IBM PS/2 and Apple II microcomputers. Most of the programs in this book should work without modification on any microcomputer, minicomputer or mainframe computer that uses the Pascal or Turbo Pascal programming language.

I would like to thank the many mathematicians and computer scientists who have used computers to discover and report on new number theory events and topics. I would also like to thank the many educators and mathematicians who have identified new terms and written about new discoveries and techniques in the field of number theory. Only through these people's works can I keep up-to-date with the ever-growing area of computerized number theory. I am particularly grateful to readers who commented on my earlier book on computerized number theory. Many of their ideas and comments were incorporated in this book. I would also like to thank Linda King for drawing many of the line illustrations and Susan Spencer for drawing most of the computer images that appear throughout the book.

I hope this book succeeds in helping readers learn about computers or solving their number theory problems.

Donald D. Spencer
Ormond Beach, Florida

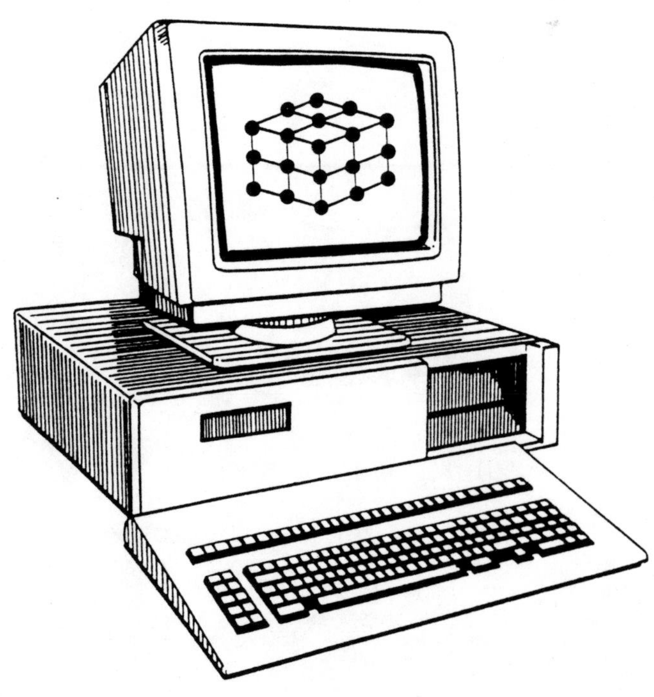

Amazing! It would take five thousand mathematicians five thousand years to make a mistake like that!

CONTENTS

PREFACE 5

CHAPTER 1 MEETING THE MICROCOMPUTER 15
- Preview 15
- 1.1 Types Of Computers 16
- 1.2 The Microcomputer System 17
- 1.3 Controlling The Microcomputer 23
- 1.4 A Glimpse Of Computerized Number Theory 23
- 1.5 Number Theory Problems For Microcomputer Solution 28
- 1.6 Representation Of Numbers In A Microcomputer 29
- 1.7 Designing Number Theory Algorithms 29
- 1.8 About The Rest Of This Book 33
- Review Exercises 33

CHAPTER 2 PRIME NUMBERS 35
- Preview 35
- 2.1 The Infinitude Of Primes 36
- 2.2 The Sieve Of Eratosthenes 42
- 2.3 Wilson's Theorem 44
- 2.4 Rare Prime Decades 46
- 2.5 Mersenne Numbers 47
- 2.6 Fermat Numbers 51
- 2.7 Twin Primes 51
- 2.8 Prime Polynomials 53
- 2.9 Emirps 54
- Review Exercises 54

CHAPTER 3 INTERESTING NUMBERS 57
- Preview 57
- 3.1 Perfect Numbers 58
- 3.2 Multiperfect Numbers 61
- 3.3 Amicable Numbers 62
- 3.4 Armstrong Numbers 65
- 3.5 Tetranacci Numbers 66
- 3.6 Lucas Numbers 66
- 3.7 Tribonacci Numbers 68
- 3.8 Lucky Numbers 70
- 3.9 Tetrahedral Numbers 72
- 3.10 Abundant And Deficient Numbers 73
- 3.11 Triangular Numbers 74
- 3.12 Rectangular Numbers 76
- 3.13 Square Numbers 77
- 3.14 Star Numbers 79
- 3.15 Powerful Numbers 80
- Review Exercises 81

CHAPTER 4 NUMBERS WITH SPECIAL PROPERTIES 83
Preview 83
4.1 Wraparound Numbers 84
4.2 Integer ABCD 86
4.3 Non-Repeating Digits 86
4.4 Integer Sums 88
4.5 Sum And Product 89
4.6 Unusual Multiplication 90
4.7 Special Four-Digit Numbers 91
4.8 Four Integers 92
4.9 Sum Of Two Squares 93
4.10 Interesting Identities 94
4.11 Powers Of Digits 95
4.12 Multigrades 96
4.13 Peculiar Property 97
4.14 Consecutive Cubes 98
4.15 Sum Of Three Cubes 101
4.16 Sum Of Factorials 102
Review Exercises 102

CHAPTER 5 FACTORING 105
Preview 105
5.1 Prime Factorization 105
5.2 Greatest Common Divisor 109
5.3 Least Common Multiple 112
5.4 Finding The GCD And LCM Of Several Numbers 113
Review Exercises 115

CHAPTER 6 FIBONACCI NUMBERS 117
Preview 117
6.1 Mathematical Patterns 118
6.2 The Fibonacci Sequence 119
6.3 Fibonacci Numbers And Primes 126
6.4 The Golden Ratio 127
6.5 Fibonacci Sorting 129
Review Exercises 131

CHAPTER 7 π IN THE COMPUTER AGE 133
Preview 133
7.1 Early Methods Of Computing π 135
7.2 Computer Calculations Of π 140
7.3 Is π Patternless? 142
7.4 Archimedes' Polygon Program 145
7.5 Leibniz Series Program 147
7.6 Wallis Series Program 148
7.7 Converging Fractions Program 149
7.8 Series Of Fractions Program 150
Review Exercises 151

CHAPTER 8 MAGIC SQUARES 155
Preview 155
8.1 Introduction 155
8.2 How To Make Magic Squares 158
8.3 Odd-Cell Magic Squares 158
8.4 Even-Cell Magic Squares 163
8.5 What Numbers Will Magic Squares Add Up To? 167
8.6 Magic Squares Starting With Numbers Other Than One 167
8.7 Multiplication Magic Squares 170
8.8 Geometric Magic Squares 173
8.9 Other Interesting Magic Squares 176
8.10 Heterosquares 177
8.11 Talisman Squares 178
Review Exercises 178

CHAPTER 9 NUMBER SYSTEMS 181
Preview 181
9.1 Introduction 181
9.2 Binary Numbers 183
9.3 Hexadecimal Numbers 184
9.4 Octal Numbers 186
9.5 Number Base Conversion Program 188
Review Exercises 190

CHAPTER 10 MODULAR ARITHMETIC 193
Preview 193
10.1 Introduction To Modular Arithmetic 194
10.2 Clock Arithmetic 195
10.3 Casting Out Nines Technique 198
10.4 Chinese Remainder Theorem 200
Review Exercises 202

CHAPTER 11 POTPOURRI 205
Preview 205
11.1 Mind Reading Tricks 206
11.2 Sailors And Coconuts 208
11.3 Diophantine Problems 209
11.4 Map Coloring Problems 210
11.5 Pythagorean Triples 212
11.6 Reciprocal Triples 215
11.7 Pascal's Triangle 216
11.8 Crossing The River 219
11.9 Fermat's And Goldbach's Conjectures 220
11.10 Palindrome Conjecture 222
11.11 Hard Problems 224
Review Exercises 226

CHAPTER 12 AN INTRODUCTION TO PASCAL 229
Preview 229
12.1 Getting Started In Pascal 231
12.2 Reading And Writing 238
12.3 Performing Calculations 247
12.4 Conditional Statements 256
12.5 Loop Control 261
12.6 Subprograms And Procedures 269
12.7 Functions 276
12.8 Recursion 281
12.9 Arrays 283
12.10 Strings 287
12.11 Enumerated Type 289
12.12 Further Pascal 290
12.13 Sample Pascal Programs 293
Review Exercises 297

GLOSSARY 301

LIST OF PROGRAMS 305

A CHRONOLOGY OF IMPORTANT NUMBER THEORY AND COMPUTER EVENTS 307

ANSWERS TO SELECTED EXERCISES 313

INDEX 317

INVITATION TO NUMBER THEORY WITH PASCAL

This number theory problem was my life's work. I planned to devote my remaining years to it. A microcomputer just solved the problem in 10 seconds

1

MEETING THE MICROCOMPUTER

PREVIEW

Computers and mathematicians have been closely related since the very first implementation of computers for problem solving. One of the primary reasons computers developed so rapidly in the 1970s was their use in solving complex mathematical and scientific problems associated with the early space program. The speed and accuracy with which the computer performs complex mathematical computations make it a necessary tool for today's mathematician and scientist.

With the availability of microcomputers, mathematics students have the opportunity to explore topics in number theory that have previously been ignored or treated superficially. The exploration of problems in number theory permits a further understanding of problem solving with microcomputers. In many ways, number theory is an ideal subject for microcomputer use. The microcomputer need only be used for computational purposes, using relatively simple programs, to make a significant contribution to a mathematics class.

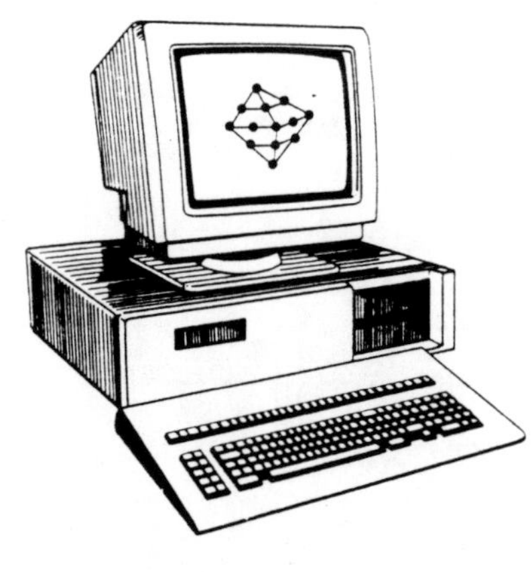

After you finish this chapter, you should be able to:

1. Identify the three main types of computers.
2. Describe some of the components of a microcomputer system.
3. Identify several number theory applications where computers are used.
4. Describe how errors may be introduced into number theory calculations.
5. List the basic steps in program development.

Minicomputers are smaller than mainframe systems but are larger than microcomputers.

1.1 TYPES OF COMPUTERS

The three types of modern computers are the mainframe computer, the minicomputer, and the microcomputer.

The largest, most costly computer is called a mainframe computer. Costing hundreds of thousands of dollars, a mainframe computer is usually housed in its own room and sometimes even in its own building! Mainframe computers are used mainly by large companies, by universities, and by government agencies. A supercomputer is a very large-scale mainframe computer that acts as a big number-crunching machine. It can process great quantities of data extremely fast and can do, in several hours, the work that normally takes weeks on conventional large mainframe computers. To qualify as a supercomputer, a machine must perform more than 20 million computations per second.

Much smaller is the medium-sized computer called a minicomputer. A minicomputer may cost tens of thousands of dollars, but it is small enough to easily fit in a small room. Unfortunately though, it is too large to sit on the user's desk. Minicomputers are used by small and medium-sized companies, by many schools, and by government agencies. They are used where a large amount of processing is required, but where a mainframe computer is not needed.

The smallest, most common, and least costly computer is the microcomputer (Figure 1-1). A microcomputer may cost anywhere from several thousand dollars to under 100 dollars. Besides low cost, a popular feature of microcomputers is their small size.

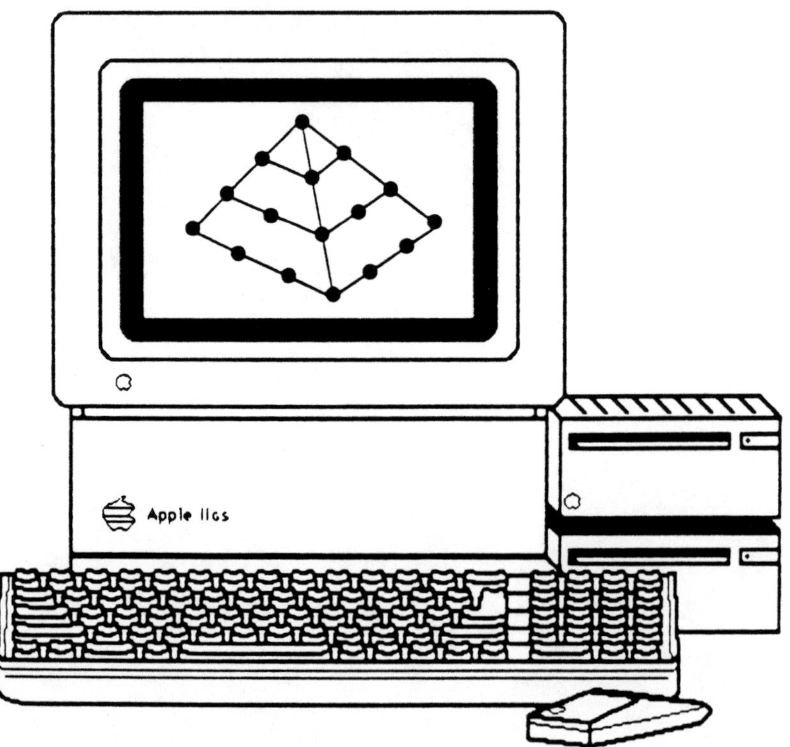

Figure 1-1
A microcomputer is a simple, lightweight, inexpensive computer system.

Microcomputers are used by schools, by businesses and by individual households. A microcomputer system contains the same types of equipment of larger computer systems but it is scaled down in size, speed, and price.

In addition to being smaller and far less costly than larger systems, microcomputers are easier to program, operate, and maintain. Because of their physical simplicity, they are less likely to break down than large systems, and repairs are easier to make. The user of a mainframe computer system must maintain a staff of operation and programming experts. In contrast, almost anyone can operate a microcomputer. In short, the basic problems of running a computer have not changed, but operations have become easier and more manageable with microcomputers.

All three types of computers can do the same kinds of jobs. However, larger computers can do several different jobs at the same time, and they can do each job more quickly.

1.2 THE MICROCOMPUTER SYSTEM

A microcomputer system consists of electronic and mechanical devices that can be directed, or programmed, to react in various ways, depending upon the needs and desires of the users. The electronic and mechanical devices (the physical pieces) are known as the **hardware. Software** refers to any directions or instructions that control the hardware. Hardware and software working together make up the total microcomputer system.

A microcomputer system is often composed of a keyboard/central processing unit, a visual display, a disk storage device, and a printer.

Keyboard And Central Processing Unit

Instructions and data are entered into the computer system through the keyboard. The microcomputer keyboard is similar to a standard typewriter keyboard. It is made up of keys that represent letters of the alphabet, numbers, and special characters, such as +, –, *, and $. In addition to these familiar characters, the microcomputer keyboard usually includes special keys that instruct the microcomputer to perform special tasks.

In many microcomputer systems, the central processing unit is included in the same housing as the keyboard. The central processing unit (CPU) is the most complex and powerful part of a microcomputer system. Like an engine in a car, the CPU is what makes the microcomputer system go. The CPU is made up of two parts:

1. Arithmetic/logic unit, which performs all calculations
2. Control unit, which coordinates the operations of the entire microcomputer system

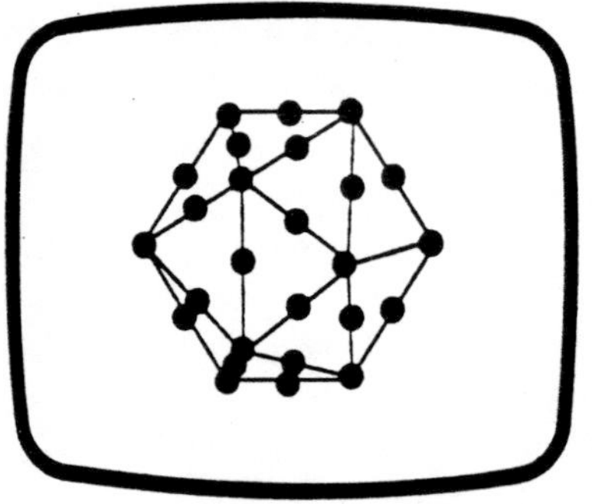

SQUARES AND CUBES

$3^2 + 4^2 = 5^2$
$5^2 + 12^2 = 13^2$
$47^2 + 52^2 = 17^3$

Figure 1-2
Most visual display devices are able to display both alphanumeric and graphic data.

The **arithmetic/logic unit** controls all operations of arithmetic and logic, and performs additions, subtractions, multiplications, and divisions on numerical data as directed by the control unit. The arithmetic/logic unit also can compare two numbers to determine whether the numbers are equal or unequal, as well as compare alphabetic information to see whether it is the same or different.

The **control unit** performs the most vital function in the CPU. It's the "brains behind the operation," where all program steps are interpreted and instructions are issued to carry out the programs. The control unit directs the overall functioning of the other units of the microcomputer and regulates data flow. When the microcomputer is operating under program control, the control unit brings in data, as required, from the input devices and directs the routing of results to the required output devices. This unit is similar to a central telephone switchboard; control is effected through the wires or circuit boards that connect all parts of the system to the central control board.

Visual Displays

Visual Displays are widely used to communicate information to computer users (Figure 1-2). Display devices enable humans to see what instructions are being given to the computer, to enter data, and to see the output generated by the microcomputer.

PASCAL'S TRIANGLE

1
1 1
1 2 1
1 3 3 1
1 4 6 4 1
1 5 10 10 5 1
1 6 15 20 15 6 1
1 7 21 35 35 21 7 1
1 8 28 56 70 56 28 8 1
1 9 36 84 126 126 84 36 9 1
1 10 45 120 210 252 210 120 45 10 1
1 11 55 165 330 462 462 330 165 55 11 1
1 12 66 220 495 792 924 792 495 220 66 12 1

A mouse is used to move the cursor on the computer's display screen.

The visual display is the most useful microcomputer output device, primarily because it is fast, quiet, and costs less than a printing device. The two most common types of terminals used in microcomputer systems are the video monitor (color or black and white) and the color television receiver. A video monitor is made specifically for information display and is similar to a home television set, but it does not include the electronics for receiving signals from a distant television station. A television receiver used with computers is simply a modified home television set.

With the development of portable microcomputers, more compact displays have become necessary. Several manufacturers are now producing flat-panel displays, thin glass-panel displays containing gases, liquid-crystal displays, or electroluminescent substances that glow when electrically stimulated. The advantages of flat-panel displays are high daylight visibility, slimness of the monitor, shock resistance, low power consumption, and longer life expectancy.

Associated with visual displays are several devices that are used to control the display cursor. Joysticks, paddles, and mice are the most popular devices used with microcomputers. A **joystick** can be tilted in any direction to control the cursor on the screen. A **paddle**, usually held in the hand, makes the cursor move either up or down, right or left. A **mouse** is attached to a computer by a long cable and is rolled along a flat surface by hand. As it rolls, the mouse controls the movement of a cursor on the microcomputer's display. A mouse, paddle, or joystick usually has a button for giving commands to the computer.

Storage Devices

Storage is actually an electronic file in which instructions and data are placed until needed. When data comes into a microcomputer through an input unit, such as a keyboard, they are converted to binary and placed in storage. The data remain there until called for by the microcomputer's control unit.

Just as you might classify the capacity of a metal file cabinet in terms of the amount of paper files it can contain, microcomputer storage is classified as to the amount of data it can hold. Some microcomputers have very small storage capacities - the equivalent of what could be printed on a few sheets of paper. Mainframe computers may be able to store the equivalent of thousands, perhaps millions, of pages of data.

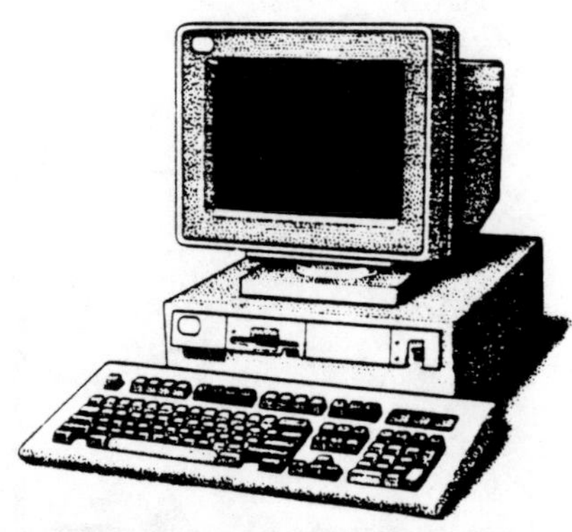

IBM Personal System/2 Color Display provides high-quality graphics and the capability of choosing up to 256 colors for any given task.

Microcomputer storage is divided into two classes: main storage and auxiliary storage. **Main storage** is an extension of the central processing unit (CPU) and is directly accessible to it. **Auxiliary storage** is used to supplement the capacity of main storage.

The major reason for the distinction between main and auxiliary storage is cost in relation to performance and capacity. Main

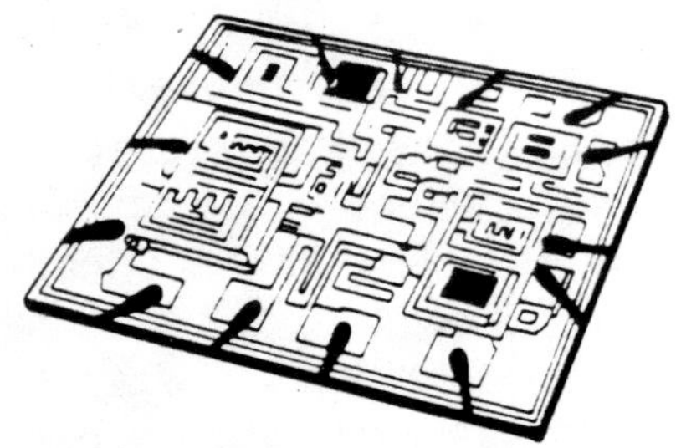

Figure 1-3
Memory chips can store more than one million bits of information.

storage must provide very fast performance and is much more costly per unit of capacity than auxiliary storage devices. Auxiliary storage must provide massive capacity for large data files - often millions or hundreds of millions of characters. However, auxiliary storage need not perform as rapidly as main storage.

Main Storage The main storage, or memory, of most microcomputers consists of microminiature semiconductor circuits. Thousands, sometimes hundreds of thousands, of semiconductor storage circuits are etched on circuit chips no larger than a matchhead. Popular memory chips contain 32, 64, 256 thousand bits (eight bits represent one byte or character). IBM Corporation, Bell Laboratories, and other companies have announced one million bit memory chips. By the turn of the century, memory chips will probably contain billions of bits (Figure 1-3).

The growth of semiconductor storage and microminiature technology has resulted in the development of two basic types of **semiconductor storage**: **random access memory (RAM)** and **read-only memory (ROM).**

RAM stores user programs and data during processing. Each storage location can be directly accessed (read or stored) in the same length of time regardless of its location in storage. Using RAM is like working on scratch paper. Unless you give the microcomputer specific instructions to save what RAM contains, the contents are lost when the microcomputer is turned off.

Figure 1-4
(Left) A 13 cm (5¼ in) diameter diskette. Diskettes are used to store programs and data. (Right) A floppy disk unit is used to read data from and write data to a diskette.

ROM is used in applications where the programs or data are not to be changed. In effect, ROM is "hard-wired" into the microcomputer when it is manufactured. The contents of ROM cannot be changed and are not lost when the power is turned off.

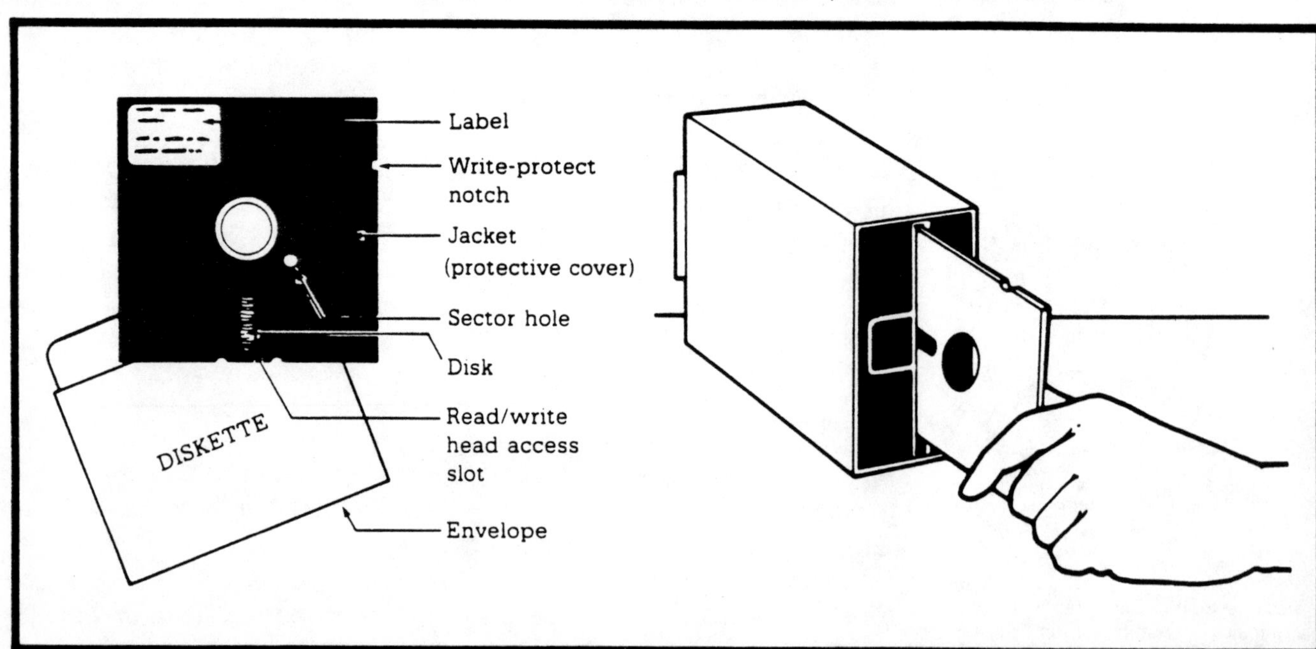

The 3½ inch diameter diskette is small enough to fit in a shirt pocket.

Auxiliary Storage Popular auxiliary storage devices for use with microcomputers are floppy disk units (Figure 1-4) and hard disks. **Floppy disks**, also called **diskettes** and **flexible disks**, are popular media for microcomputer software. The floppy disk is a compact, flexible, magnetic-oxide coated mylar disk. It comes in three popular diameters:

- 8 inches (20 centimeters)
- 5¼ inches (13 centimeters)
- 3½ inches (9 centimeters)

All diskettes are enclosed in jackets for protection, with a slot for access by the disk's reading and writing mechanism. While the disk is rotating, information is recorded in digital fashion on the disk's magnetic surface. A diskette will hold anywhere from 130,000 characters to well over one million characters. The amount depends to some extent on the price of the disk drive.

The read/write head of floppy disk units occasionally comes in contact with debris that may cause it to "crash" and lose data. Special devices have been developed to overcome this problem. One such device is the **hard disk drive**. Hard disks are sealed modules that contain both a disk and a read/write head mechanism. Since the disk module is sealed, head-to-disk alignment problems are eliminated, and the risk of exposing the recording surface to airborne contaminants is reduced. Hard disks have the highest storage capacity, lowest cost per stored byte, and lowest cost of movable-head rigid magnetic disk drives.

Printing Devices

Among the most useful output devices for use with microcomputers is the printer (Figure 1-5). Printers enable you to make paper copies of stored information, and they come in all shapes, sizes, and price ranges. The two main classifications of printers are impact and non-impact. As the name implies, **impact printers** have a type element that produces a printed image by striking the paper. The element usually strikes a ribbon on its way to the paper. Impact printers use

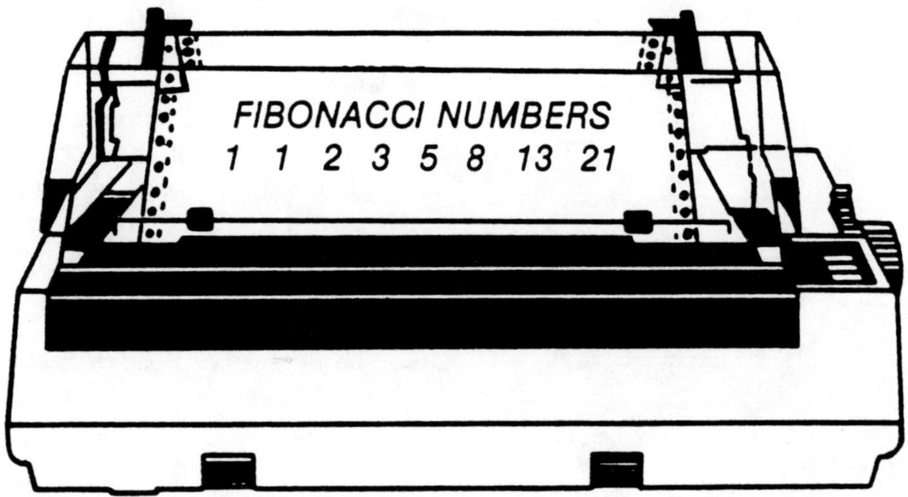

Figure 1-5
Printers are useds to produce a hard copy printout of a program listing or program output.

NUMBER THEORY TRIVIA

Analytic number theory is a branch of number theory concerned with the interaction of analysis and number theory. Analysis can be used to prove certain properties of ordinary and algebraic integers and to establish quantitative results. Similarly, arithmetic properties can be used to analyze and shed light on analytic questions. Among the earliest problems in number theory to which any sort of systematic study was brought to bear is the class of problems known as Diophantine equations, named after the Greek mathematician Diophantus of Alexandria who lived about 250 A.D.

typefaces mounted on elements such as balls, thimbles, wheels, drums, or chains. The type on these machines is solid, produces sharp impressions, and is used in systems where letter-quality printing is required.

A low-cost impact printer is the **matrix printer**, also called the **dot matrix printer**. This machine uses a movable print head that consists of a matrix of small tubes, each confining a fine wire or needle. The needles are fired individually by electrical solenoids (devices that convert electrical energy into mechanical energy) to produce a dot matrix that varies in size and shape. The higher the density of the dots forming the matrix, the better the printing. Several manufacturers now produce dot matrix printers with high density matrices that nearly equal the print quality of solid type.

The **daisy wheel printer** uses a petal-shaped wheel with the tip of each petal containing a single character. The print wheel revolves to bring each character into print position; then a single solenoid fires to print the character. Daisy wheels are slower than matrix printers, but produce solid, higher-quality characters and come in a variety of typefaces that can be changed easily.

Non-impact printers produce a printed image without striking the paper. These machines operate at high speeds without the clatter associated with impact printers, but they produce only one copy at a time. Most use special paper that is more expensive than that used with other printers.

The **ink jet printer** shoots a steady stream of tiny ink droplets toward the paper. The computer controls which droplets are charged with electricity. Most of the droplets are magnetically attracted away from the paper, while a few drops hit the paper to form symbols. The resulting printed characters are of a very high quality.

Thermal printers, another common non-impact printer, operate by selectively activating a series of printing elements to form each character. These elements "burn" the character into specially treated heat-sensitive paper.

The **laser printer** uses a microthin beam of light to "draw" each character or graphic. It is capable of extremely high resolution and can print several thousand copies per minute. A few years ago, the typical laser printer was hooked up to a mainframe computer and cost upwards of a quarter of a million dollars. Now fast, efficient laser printers hook up to microcomputers and cost between $2000 and $6000.

Laser printers, matrix printers, ink jet printers, and thermal printers can produce either color or black and white hard copy.

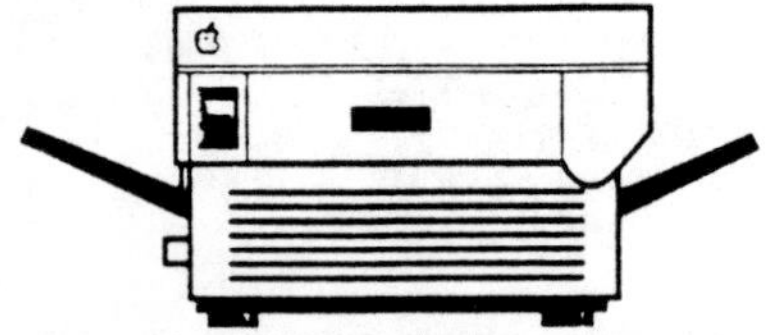

A laser printer produces a high quality printout.

1.3 CONTROLLING THE MICROCOMPUTER

You have already been introduced to the major hardware components of all microcomputer systems. Now you will learn how microcomputer systems are used to solve problems.

A **program** is a set of instructions that tells the microcomputer how to perform a specific calculation or operation. The user communicates a problem to the microcomputer via the program. Each instruction in a program is a direction to the microcomputer to perform some simple operation, such as adding two numbers or printing a line of results. The microcomputer follows the sequence of instructions in a program without deviating and is thus able to solve problems without human intervention.

A program may be written in one of several available programming languages. One such language is **machine language**, the basic language of the microcomputer; however, machine language is seldom used because it is very difficult. A language on the same level as machine language, but one that consists of easy-to-use symbolic codes rather than numbers, is called **assembly language**. These symbolic codes are employed to represent microcomputer instructions, storage locations, and so on.

Other easy-to-use programming languages are called **high-level languages**. These languages let users write programs in terms that describe problems and their solutions, rather than in terms relevant to the internal operation of microcomputers. Several high-level languages are in use. Two of the most popular are BASIC and Pascal.

Before a program written in a high-level language can be executed on a computer, it must be translated into machine language. A **compiler** performs a complete translation of the set of instructions before the program is actually executed, and an **interpreter** converts each instruction to machine language as the program is executed. Compilers and interpreters are also programs, whose sole purpose is to perform the translation process.

Euclid (Circa 300-275 B.C.) has been called "The Father of Geometry." He founded and taught the school of mathematics in Alexandria and, while there, wrote several books. Euclid's writings, in accordance with the practice of the time, were written on rolls of parchment or papyrus. Although Euclid may not be called the greatest geometrician, he certainly was one of the greatest organizers and compilers of materials on that subject. There are thirteen books in his Elements, the earliest textbook on geometry still extant. His first book deals with plane geometry. The others concern themselves with ratio and proportion, solid geometry, polygons, circles and number theory.

1.4 A GLIMPSE OF COMPUTERIZED NUMBER THEORY

Number theory is a field of study that has excited the interest of mathematicians for centuries. It is primarily concerned with properties of the positive integers (1, 2, 3, . . .). It is one of the two oldest branches of mathematics, geometry being the other. But whereas the classical "circle and triangle" geometry of Euclid is essentially dead as a research science, number theory still contains a large body of challenging and unsolved problems. Indeed, some of its most intriguing problems go back to Euclid's time.

Pythagoras was a teacher of arithmetic and number theory. Pythagoras and his followers studied numbers as geometric arrangements of points, such as the triangular numbers 1, 3, 6, 10, 15, 21, 28, 36, 45, and so on.

In early western and oriental cultures, whole numbers were originally of interest for ritual and computational purposes. The study of their properties, however, seems to have begun with the Pythagoreans, a school of scholars led by Pythagoras (572 B.C.), of Pythagorean Theorem fame. They felt that certain whole numbers had mystical significance.

Over the following two centuries, mathematicians of the time considered various types of whole numbers: for example, prime numbers (numbers that have exactly two divisors), perfect numbers (numbers that are equal to the sum of their proper divisors), and deficient numbers (numbers that exceed the sum of their proper divisors). Euclid (300 B.C.) was able to characterize all even perfect numbers, but to this day it is not known whether or not any odd perfect numbers exist.

Number theory has many practical applications in engineering and physics, and has many uses in proving theorems in other fields of mathematics. An engineer who designs a gear train must use a form of number theory, and so must a physicist who undertakes an explanation of the interactions between atoms and radiation.

Many of the problems in the field of number theory are simple to state but difficult to solve. In recent years, computers have been used to aid mathematicians in the field of number theory. Several examples using the computer to solve number theory problems will be considered in this section. Many more examples are given throughout the book.

Karl Gauss (1777-1855), one of the greatest mathematicians of all time, called **number theory** "the queen of mathematics." Gauss also called mathematics "the queen of the sciences."

ARCHIMEDES
(287-212 B.C.)

The following problem appeared in a paper written by Archimedes. The sun had a herd of bulls and cows, all of which were either white, grey, dun, or piebald: the number of piebald bulls was less than the number of white bulls by 5/6ths of the number of grey bulls, it was less than the number of grey bulls by 9/20ths of the number of dun bulls, and it was less than the number of dun bulls by 13/42nds of the number of white bulls; the number of white cows was 7/12ths of the number of grey cattle (bulls and cows), the number of grey cows was 9/20ths of the number of dun cattle, the number of dun cows was 11/30ths of the number of piebald cattle, and the number of piebald cows was 13/42nds of the number of white cattle. The problem was to find the composition of the herd. The problem is indeterminate, but the solution in lowest integers is

white bulls	10,366,482
grey bulls	7,460,514
dun bulls	7,358,060
piebald bulls	4,149,387
white cows	7,206,360
grey cows	4,893,246
dun cows	3,515,820
piebald cows ...	5,439,213

In the seventeenth century, Marin Mersenne (1588-1648) acted as a kind of mathematics broadcaster. He corresponded with famous mathematicians of his day, and his letters spread news about inquiries and discoveries.

Factoring Large Numbers

Mathematicians at the Sandia National Laboratories in Albuquerque, New Mexico used a Cray 1S supercomputer to factor a 69-digit number, the last unfactored number in the list compiled by the seventeenth-century French mathematician Marin Mersenne. The number was found to be the product of three primes, 21 digits, 23 digits, and 26 digits long. The final Mersenne number had lived up to its reputation by proving to be especially difficult to factor even with all the new methods and technology. The process took 32 hours and 12 minutes of computer time.

The number $2^{251} - 1$ is

132686104398972053177608575506090561429353935989033525802891469459697

and has the factors

178230287214063289511,
61676821986952575 01367,

and

12070396178249893039969681.

Seymour Cray invented the world's first popular supercomputer, the Cray 1.

The factoring of very large numbers has, in the past year or so, become very significant because of its use in such two-key encoding systems as electronic funds transfers, "smart" credit cards, and automatic bank teller machines. The widely used RSA cryptographic system employs multidigit numbers to encode secret information, including electronic funds transfers and military messages. When RSA was introduced its inventor suggested using 80-digit numbers, assuming they were too large to be factored. However, the Sandia team has already gone on to factor the extremely difficult 71-digit number consisting of 71 ones. This was accomplished in 6.45 hours of computer time.

By piecing together the output of hundreds of computers on three continents, a team of mathematicians succeeded in solving a computational problem that had defied all previous efforts. On October 11, 1988, the last sequence of numbers required for the solution popped up in a computer laboratory in California, and news of the triumph was flashed to collaborators around the world. The team had successfully factored a number 100 digits long into two prime factors which are respectively 41 digits and 60 digits long.

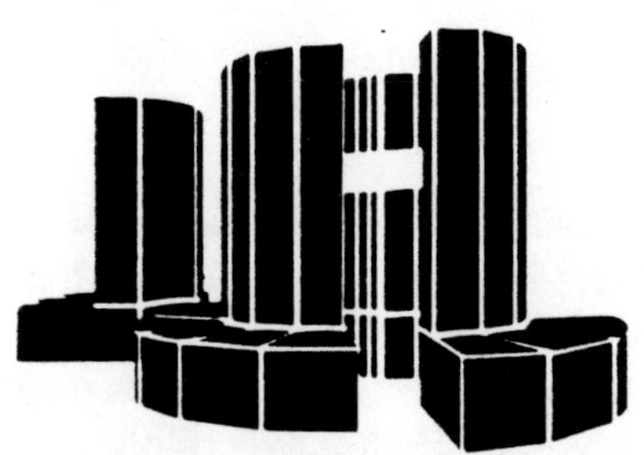

Cray supercomputers have been used by mathematicians to produce many new discoveries in number theory.

Several of the most secure cipher systems invented in the past decade are based on the fact that large numbers are extremely difficult to factor, even using the most powerful computers for long periods of time. The accomplishment of factoring a 100-digit number may possibly prompt cryptographers to reconsider their assumptions about cipher security.

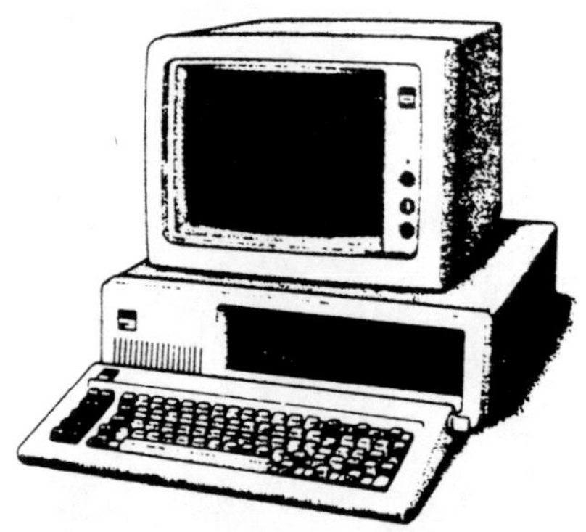

Michael Friedman used an IBM Personal Computer to demonstrate that there are no odd perfect numbers below 10^{79} that have eight prime divisors.

Is There An Odd Perfect Number?

Michael Friedman, now a student at MIT, was a high school senior from Brooklyn when he won third place in a Westinghouse Science Talent Search. For his award-winning project he didn't want to dirty his hands with brine shrimp, fruit flies, or flatworms. And he didn't want to tackle just any age-old theoretical question. No, he chose to confront a problem so old that it could well be the oldest unsolved problem in mathematics, a problem that confounded the ancient Greeks and everyone since: Is there an odd perfect number?

Pythagoras and his cronies saw perfection in any whole number that equals the sum of all its divisors (except the number itself). The first perfect number is 6. It's evenly divisible by 1, 2, and 3, and it's also the sum of 1, 2, and 3. The second perfect number is 28. Its divisors are 1, 2, 4, 7 and 14, and they add up to 28. That much the Greeks knew, but try as they did, they could not find an odd perfect number.

Friedman learned that number theorists, frustrated by the intractability of the problem, have proved all sorts of things about what an odd perfect number must be like if one exists. It must be evenly divisible by at least eight different prime numbers, of which the largest must exceed 300,000 and the second largest must exceed 1,000. If an odd perfect number is not divisible by 3, it must be divisible by at least 11 different prime numbers. Moreover, an odd perfect number must leave a remainder of 1 when divided by 12 and a remainder of 9 when divided by 36.

With an IBM Personal Computer and a list of constraints, Michael demonstrated that there are no odd perfect numbers below 10^{79} that have eight prime divisors (which is the minimum number of prime divisors an odd perfect number could have).

Slicing π Into Millions

Pi, from the Greek letter π, is the ratio of the circumference of a circle to its diameter, a constant discovered in ancient times. Its first few digits - 3.14159 - are familiar to most people. Mathematicians originally calculated π roughly, by measuring actual circles, but soon developed much more precise, purely arithmetical, methods for determining π to any desired accuracy.

Francois Viete made important contributions to arithmetic, algebra, trigonometry and geometry. He also introduced a number of new words into mathematical terminology, some of which, such as *negative* and *coefficient*, have survived. His attack on π resulted in the first analytical expression giving π as an infinite sequence of algebraic operations.

The first computer to tackle π was ENIAC (the first large-scale electronic digital computer), which calculated its value out to 2037 places. This large computer took several hours to complete the job. During the next four decades many computers were used to compute π to much greater accuracy. In 1989, two Columbia University research scientist brothers, David and Gregory Chudnovsky, computed π to 480 million decimal places.

LARGEST PRIME

- 65,000 DIGITS LONG
- DISCOVERED BY DAVID SLOWINSKI

Biggest Prime Number

A prime number is a number that can be divided evenly only by 1 or by itself. Known primes range from small numbers like 2, 3, 5, 7, 11, and 13 to a large number 65,050 digits long. This large number was found by David Slowinski, a computer scientist at Cray Research in Chippewa Falls, Wisconsin. Slowinski used a Cray X-MP 24 supercomputer to produce this number which fills 14 pages of computer printout. It is far too large to have any meaning in the physical world. For comparison, a mere 125-digit number is greater than the number of subatomic particles in the known universe.

For hundreds of years, primality was a mathematical curiosity but was not of much practical importance. The computer changed all that. Prime numbers now play a key role in the fast Fourier transform algorithms (step-by-step procedures) used for digital signal processing, in generating random numbers used for computer simulation studies of physical phenomena, and in cryptography. These new uses stimulated research into finding methods for generating prime numbers and methods for testing whether a number is prime.

Large prime numbers, those with more than 20 or 30 digits, are of particular importance in computer algorithms, and especially in cryptography. Many modern encryption techniques rely on large prime numbers as the key ingredients in making encoded messages difficult to dicipher.

Undoubtedly, the search for new and larger prime numbers will continue by mathematicians and computer scientists who have access to the occasional idle moments of large, fast supercomputers.

Sum And Product

Some numbers have the property that the sum and product of the digits are equal. For example, 22 has the property because $2 + 2 = 2 \times 2$, but 57 does not because $5 + 7 \neq 5 \times 7$. To see if any three-digit numbers have the property, we must test every number of the form ABC to see whether $A + B + C = A \times B \times C$. Performing the test by hand calculating methods would be extremely consuming. By using a microcomputer and a short program, you can produce just one set of digits: 1, 2, and 3. All possible permutations of these digits produce six solutions to the problem: 123, 132, 321, 312, 231, and 213.

Joseph Louis Lagrange (1736-1813), a mathematician who excelled in all fields of analysis and number theory. By 1761 he was recognized as the greatest living mathematician.

By using a microcomputer, you can easily find the four-digit numbers that have this property. That analysis would again produce just one set of digits: 1, 1, 2, and 4, whose permutations generate twelve solutions. An analysis of five-digit numbers gets a little more interesting, with three sets of digits and forty solutions.

Leonhard Euler was born in 1707 at Basel, Switzerland. He was the most eminent and influential mathematician of the 18th century and he was by far the most prolific mathematician of all time. His discoveries in mathematics and in many fields of science are so numerous that his collected work will eventually fill about 80 volumes. All of his life, Euler worked intensively on problems in number theory.

An Interesting Number

It is extremely interesting to ponder the fact that 26 is the only number (less than 10 quadrillion, at least) to be "squeezed" between a square, 25, and a cube, 27. That's a very nice property. But there is little that one can do to verify the statement without a computer. With a computer, however, one might even extend the search beyond 10 quadrillion.

1.5 NUMBER THEORY PROBLEMS FOR MICROCOMPUTER SOLUTION

This section includes several problems that are representative of number problems that can be solved on a microcomputer. Try solving one of these problems without using a microcomputer.

Square Numbers

The square of 12 is 144. The digits of 144 reversed are 441 - another perfect square. Moreover, the square root of 441 is 21, the reverse of 12. Program a microcomputer to find other numbers with this property.

Consecutive Digit Primes

23 is the first prime number consisting of two consecutive digits, 2 and 3. Such primes are quite rare. Other two digit primes are 67 and 89. Use a microcomputer to determine that no 3-digit primes consisting of 3 consecutive numbers exist. Also, program a microcomputer to determine that only one such 4-digit prime number exists; it is 4567. Determine if the next such prime is the 8-digit number 23,456,789.

The brilliant Indian mathematician Srinivasa Ramanujan is pictured on this postage stamp. He developed many results in number theory. He specialized in the study of numbers and knew their characteristics in the same way that a baseball fan might know a vast number of statistics about the game. One time a friend went to visit him in a taxi having the number 1,729. When the friend mentioned this number, Ramanujan immediately replied: "1,729 is a very interesting number; it is the smallest number expressible as the sum of two cubes in two *different* ways."

$$1729 = 1^3 + 12^3 = 9^3 + 10^3$$

Rare Primes

97 is the only prime number in the 90s. 113 is the only prime in the decade from 110 to 119. 127 is the only prime in the decade from 120 to 129. 149 is the only prime in the decade from 140 to 149. 181 is the only prime in the decade from 180 to 189. Use a microcomputer to determine all decades less than 5000 that contain only one prime number.

Digit Sum

371 is equal to the sum of the cubes of its digits (27 + 343 + 1 = 371). Use a microcomputer to determine other 3-digit numbers with the same property.

Divisors

60 is the smallest number with 12 divisors. Those divisors are 1, 2, 3, 4, 5, 6, 10, 12, 15, 20, 30, and 60. Use a microcomputer to determine all other 2-digit numbers with 12 divisors.

NUMBER THEORY TRIVIA

Whereas ordinary number theory is the study of the integers, in **algebraic number theory** larger collections of numbers are studied. A number is algebraic if it is a root of a polynomial equation with integral coefficients; if the highest power of the variables has coefficient 1, the number is called an algebraic integer. For instance, the square root of 2 is an algebraic integer because it satisfies the equation $x^2 - 2 = 0$.

1.6 REPRESENTATION OF NUMBERS IN A MICROCOMPUTER

When using a computing device, we must consider what errors will be introduced into calculations. In a microcomputer, there is only a limited amount of space in which to store each number, and only the most significant digits in the number will be retained; the actual number of digits is governed by the word length of the particular machine. For example, consider calculating π^2 on a machine which will store six digits. The number $\pi = 3.1415926535\ldots$ would be stored as 0.314159×10^1 (the decimal part is called the mantissa and the index of 10 the exponent). Multiplication of π by π would give, at best, $0.098695877281 \times 10^2$, which would be stored as 0.098696×10^2 and returned as 9.8696. As compared with the true value of π^2, this calculation gives an error in excess of 0.000004.

It is not always possible to store real numbers exactly in the internal representation of a microcomputer. Some values just cannot be represented in a finite number of digits. For example, in decimal:

$$1/3 = .3333333$$

Regardless of where we stop, we will not have the exact representation of 1/3. Our representation will have introduced a small, but distinct, truncation error. If we now perform computations using this truncated value the results will also be slightly in error. Microcomputers perform their computations in binary (base 2), not decimal (base 10). However, since binary is also a positional numbering system, it also suffers from the same problem.

Muhammed idn Musa al-Khowarizmi (780-850) was a Persian mathematician and astronomer who lived in Baghdad and whose name has given rise to the word **algorithm**. He wrote a book *Hisab al-jabr wal-mugabalah*, which was a compilation of rules for solving linear and quadratic equations, and has given rise to the word **algebra**.

1.7 DESIGNING NUMBER THEORY ALGORITHMS

A very important step in developing a solution to a number theory problem is the analysis of the problem. Often, questions of the following type must be asked:

- Do we know how to solve the problem on a microcomputer?
- Can the problem be solved with a microcomputer?
- Can the microcomputer in question solve the problem?
- Has the problem already been solved?
- Is the problem worth doing?

After questions of this type are answered, one can better determine if a problem should be solved by a microcomputer.

After all elements and relationships in the problem have been studied and defined in specific terms, they must be expressed as steps that the microcomputer can perform. This sequence of steps for solving a problem is called an **algorithm**. Simply stated, an algorithm is a recipe or list of instructions for doing something. More precisely defined, it is a complete procedure or plan for solving a

How did I get into this business? Well, I couldn't understand number theory in college, so I settled for this instead.

problem. An algorithm provides the logical steps the microcomputer will follow in solving the problem.

We solve most decision problems by using algorithms - by breaking the problem down into many steps. Problems for microcomputer solution must be broken down into many simple steps or instructions. The algorithms for many problems are rather simple, however algorithms for complicated scientific problems can be quite complex. The algorithm for determining if 379 is a prime number is simple: divide 379 by each of the numbers 2, 3, 4, . . ., 378. If any division results in a zero remainder, the number is not prime; otherwise it is a prime number. The algorithm for simulating the flight of an airplane is complex and involves several thousand steps.

An algorithm must have the following four characteristics to be useful. It should be (1) unambiguous, (2) precisely defined, (3) finite, and (4) effective. The following examples illustrate these characteristics.

NUMBER THEORY TRIVIA

Geometric number theory is a branch of number theory that can be developed by the use of certain geometric methods. It centers on the arithmetical theory of quadratic and higher forms and the problems of the approximation of real numbers by rational numbers.

Leonardo Fibonacci

Fibonacci Number Algorithm

Leonardo Fibonacci, a wealthy Italian merchant of the thirteenth century, introduced a set of numbers which are known as *Fibonacci Numbers*. The first fourteen of them are:

1, 1, 2, 3, 5, 8, 13, 21, 34, 55, 89, 144, 233, 377

Each number of a Fibonacci number sequence is the sum of the two numbers immediately preceding it, that is

1 + 1 = 2
1 + 2 = 3
2 + 3 = 5
3 + 5 = 8
5 + 8 = 13
8 + 13 = 21 . . .

The number sequence has practical applications in botany, electrical network theory, and other fields.

The Fibonacci number sequence is formed by the following algorithm:

1. Set A, B, and S equal to 1
2. Write down A and B
3. Replace A with B and B with S
4. Let S = A + B
5. Write down S in the sequence
6. Go back to Step 3 and proceed through Step 6 again.

Number Selection Algorithm

There are often several different algorithms for solving the same problem. Development of different methods often reflects a personal style or insight. Some people are very clever at finding algorithms which give an answer quickly. Other people tend to use familiar approaches which may require longer time. For example, let us develop an algorithm for finding the largest number in the list of numbers:

26, 114, 9, 82, 61, 155, 4, 19, 3, 183

We first scan the list and eliminate all one-digit numbers, leaving 26, 114, 82, 61, 155, 19, 183. Next we eliminate all two-digit numbers, leaving 114, 155, 183. Next we eliminate all three-digit numbers. Whoops! There is nothing left. Then we back up one step and compare the three-digit numbers with each other, as the answer is among these numbers.

Let us compare the first to the second; 155 is larger than 114, so we throw out 114. This leaves 155 and 183. Comparing these two and throwing out 155 leaves us the final answer of 183. Another way of solving the same problem would have been just to compare the first

NUMBER THEORY TRIVIA

In 1976, the Four-Color Problem was solved: every map drawn on a sheet of paper can be colored with only four colors in such a way that countries sharing a common border receive different colors. The proof made unprecedented use of computer computation; the correctness of the proof cannot be checked without the aid of a computer.

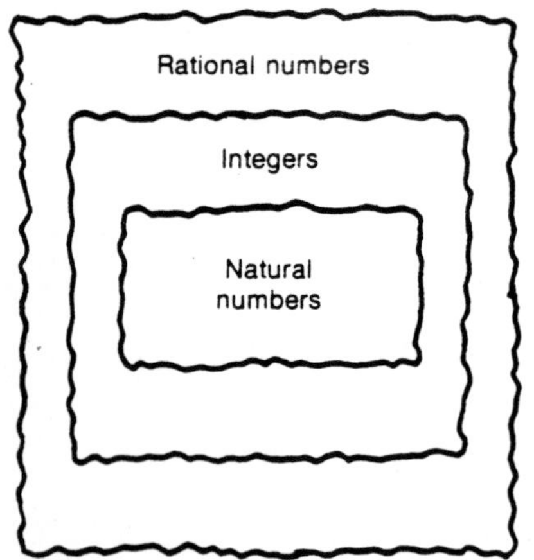

Natural numbers are the counting numbers 1, 2, 3, 4, . . .

Integers — the set of whole numbers — -3, -2, -1, 0, 1, 2, 3

Rational numbers are fractions, of the form *a*/*b*, where *a* and *b* are whole numbers.

two numbers and throw out the smaller, continuing this process until only one number, the answer, is left. For example, in the original set of numbers we would compare 26 to 114 and throw away 26; then 114 to 9 and throw away 9; then 114 to 82 and throw away 82; and so forth until we compared 115 to 183 which would determine the final answer. A number selection algorithm of this type must work for any problem of this type. For example, the algorithm must work for: 26, 148, 0, 61, 182, as well as 14, –68, 142, -1, 18.

Number Game Algorithm

Have you ever had someone ask you to think of a number and tell you several things to do with it? Then, without knowing your original number, the person is somehow able to tell you what number you ended up with. For example: "Choose a number. Add 5. Double the result. Subtract 4. Divide by 2. Subtract the number you started with." (*Slight pause.*) "Your result is 3." It is always 3, no matter what the original number was.

How does this trick work? Let's look at an algorithm that proves that the result is *always* 3.

1. Choose a number	n
2. Add five.	n + 5
3. Double the result.	2n + 10
4. Subtract four.	2n + 6
5. Divide by two.	n + 3
6. Subtract the number you started with and the result is three.	3

Now it is easy to see why the original number makes no difference in the final result.

Converting The Algorithm To A Program

Structured programming is a well-organized method for designing an algorithm. It is a set of conventions and rules that, when followed, yield programs that are easy to write, easy to test, easy to read, and easy to modify. In short, structured programming forms the basis of a program design notation. Structured programming uses a top-down approach where the entire solution is defined initially at the highest level of abstraction. Top-down design essentially means proceeding by refinement from the highest level down to the lowest level. At each level, the function to be performed is defined and then expressed in terms of functional units at a lower level, containing the process until the level is reached at which one can code the program.

After the algorithm has been developed, it must be converted into a set of instructions understandable to the microcomputer. Coding is the process of converting the steps in the algorithm to a set of instructions written in a programming language. This set of

NUMBER THEORY TRIVIA

From the development of the arithmetical calculations of finding two-thirds of any number, the Egyptians were able to determine with surprising accuracy, to solve first- and second-degree equations, and to sum arithmetic and geometric progressions.

NUMBER THEORY TRIVIA

Numerology is the use of numbers to interpret a person's character or to divine the future. It is based on the Pythagorean idea that all things can be expressed in numerical terms because they are ultimately reducible to numbers. Using a method analogous to that of the Greek and Hebrew alphabets (in which each letter also represented a number), modern numerology attaches a series of digits to an inquirer's name and date of birth and from these purports to divine the person's true nature and prospects.

instructions, called a program, will cause the microcomputer to produce the solution to the problem.

Throughout this book the Pascal programming language will be used to develop programs.

1.8 ABOUT THE REST OF THIS BOOK

Although the examples shown in Sections 1.4 show the power of the computer, they also show that the honor of finding the next largest prime number, the next perfect number, a more accurate calculation for π, or any other special number will last only until someone else is willing to devote more computer time to the problem.

The emphasis throughout this book is on microcomputer solutions to number theory problems. The program solutions included in this book are written in the Pascal programming language. If you are not familiar with this language you may wish to study Chapter 12. We will use the microcomputer to generate some special numbers as well as to solve many other interesting problems in number theory (Chapters 2 through 11). All chapters are, for the most part, independent of one another. Many readers or teachers may prefer to rearrange the chapter reading sequence or to delete one or more chapters as time and applicability dictate. The variety of topics and the independence of the chapters facilitate this flexibility.

REVIEW EXERCISES

1. Name the three major types of computers.
2. Explain the difference between a mainframe computer and a microcomputer.
3. What are some of the advantages of using a microcomputer over using a minicomputer or mainframe computer?
4. What is meant by hardware? Software?
5. What is a microcomputer system?
6. What is a central processing unit?
7. What is the purpose of the arithmetic/logic unit?
8. What unit directs the overall functioning of other units of a microcomputer?
9. What is the purpose of a keyboard?
10. What is the most useful microcomputer output device?
11. What is the name of the device that can be used to control the display cursor?
12. What is the purpose of a storage device?
13. Identify the two main classes of storage.

On October 12, 1988, Steve Jobs, a co-founder of Apple Computer Inc., announced a new computer from his new company, NeXT Inc. The NeXT Computer System offers sophisticated graphics, stereo sound and an erasable optical disk memory. The machine is designed to be used in colleges and schools.

NUMBER THEORY TRIVIA

Euclid is the only man to whom there ever came, or can ever come again, the glory of having successfully incorporated in his own writing all the essential parts of the accumulated mathematical knowledge of his time.

David E. Smith
Twentieth-century mathematician

14. What is the purpose of an auxiliary storage device?
15. Identify two basic types of main storage found in microcomputer systems.
16. What is the main storage of a microcomputer system used for?
17. Identify two popular types of auxiliary storage.
18. What output device can be used to produce a paper copy?
19. What is the purpose of a program?
20. What is the basic language of a microcomputer?
21. Give an example of a high-level programming language?
22. Briefly explain how computers have been used in solving number theory problems.
23. What type of computers were used to determine the factors of Mersenne's 69-digit number, to compute the largest prime number, and to compute π to several million decimal places?
24. What type of computer did Michael Friedman use to demonstrate his answer to the odd perfect number question?
25. David Slowinski used a Cray supercomputer to compute a prime number that contained ________________ digits.
26. Briefly explain how errors may be introduced into calculations performed on a microcomputer.
27. What is an algorithm?
28. Write an algorithm for determining if 283 is a prime number.
29. Select two consecutive odd numbers larger than 79, neither of which is divisible by 3. Multiply the numbers and subtract 79. Divide by 4 and subtract 79 again. Divide by 9 and subtract 79. Finally, subtract 79 one more time. The result will always be composite and can be factored into two factors such that the larger exceeds the smaller by 26. Write an algorithm that describes this process.
30. Select any 2-digit number. Reverse its digits and add them to the original number. Divide the sum by 0.125. Subtract the sum of the digits of the number originally chosen. The result is divisible by 87. Write an algorithm that illustrates this process.
31. What is meant by structured programming?
32. The process of converting an algorithm to a program is called ________________ .

2 PRIME NUMBERS

PREVIEW

We begin our study of computerized number theory by looking at a special subset of counting numbers, the prime numbers, which Pythagoras and other Greek mathematicians studied at length. In fact, prime numbers have been one of the key ideas studied throughout the history of mathematics.

Prime numbers are useful in analyzing problems concerning divisibility, and are interesting because of some of the special properties they possess as a class. These properties have fascinated mathematicians and others since ancient times, and the richness and beauty of the results of research in this field have been astonishing. Mathematicians and computer scientists have spent enormous amounts of time computing large prime numbers and testing whether certain large numbers are prime. It is in this area that the computer has made a spectacular contribution to number theory.

FERMAT PRIMES
3
5
17
257
65537

After you finish this chapter, you should be able to:

1. Identify prime numbers, Mersenne numbers, and twin primes.
2. See how a microcomputer can be used to produce prime numbers and twin primes.
3. Identify several techniques and formulas that can be used to generate prime numbers including trial division, sieve of Eratosthenes, Wilson's theorem, and limited prime number formulas.
4. Write programs to produce prime numbers.

NUMBER THEORY TRIVIA

The following table shows primes that are the **sum of primes**. Further there are no duplicating digits appearing in the integers involved.

5 = 2 + 3
7 = 2 + 5
41 = 5 + 7 + 29
47 = 5 + 13 + 29
61 = 2 + 59
67 = 5 + 19 + 43
89 = 5 + 23 + 61
103 = 2 + 5 + 7 + 89
401 = 5 + 7 + 389
809 = 5 + 43 + 761

2.1 THE INFINITUDE OF PRIMES

The prime numbers are the multiplicative building blocks of the number system. If a number is prime, there are no smaller natural numbers that can be multiplied to yield it as their product. The prime number 13, for example, cannot be broken down into smaller factors; only 1 × 13 is equal to 13. If a number is composite, on the other hand, it can be expressed as the product of two or more prime factors. The composite number 15 is equal to 3 × 5. Every whole number larger than 1 is either a prime or the product of a unique set of primes. The first few primes are:

2, 3, 5, 7, 11, 13, 17, 19, 23, 29, 31, 37, 41, 43, 47

A glance reveals that this sequence does not follow any simple law. In fact, the structure of the sequence of primes is extremely complicated.

The prime numbers are scarce when we consider large number ranges (see Figure 2-1). The reason for this is clear: the larger a number is, the more numerous its potential divisors, and the less likely it is to be a prime. Nevertheless, the list of prime numbers appears to be endless. In fact, Euclid proved that there is an infinite number of primes.

How can one determine whether a number is prime or composite? The most straightforward way is to divide the number to be tested by the integers in sequence: 2, 3, 4, 5, 6 and so on. If any of the divisions come out even (that is, leaves no remainder), the test number is composite and the divisor and the quotient are factors of the number. If all the integers up to the test number are tried and none of the divisions come out even, the number is prime. Actually it is not necessary to continue up to the test number; the procedure can be stopped as soon as the trial divisor exceeds the square root of the test number. The reason is that factors are always found in pairs; if a number has a factor larger than the square root, it must also have one smaller.

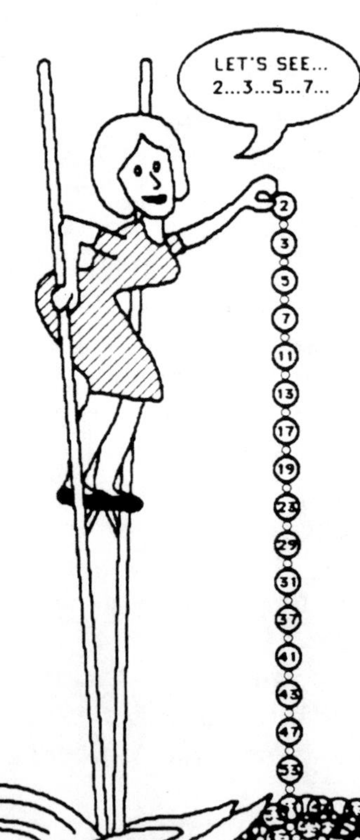

Stopping the trial division at the square root can greatly speed up a test for primality and there are other short cuts, such as deleting all the even divisors after 2.

Example 1. Is 91 a prime? $\sqrt{91}$ = 9+; by trying the numbers 1, 2, 3, 4, 5, . . ., one sees that 91 = 7 × 13.

Example 2. Is 1973 a prime? $\sqrt{1973}$ = 44+. Since no number less than or equal to 43 divides 1973, this number is prime.

In general terms, if the factors of N are 1, 2, 3, 4, . . ., N, the whole range is uncovered by dividing by:

$$1, 2, 3, 4, \ldots, \sqrt{N}$$

NUMBER THEORY TRIVIA

$217833 \times 10^{7150} + 1$ is a prime number of the form $k \times 10^n + 1$, discovered in 1985 by Harvey Dubner. It took five weeks to test it on a microcomputer system. It is a large, non-Mersenne prime that contains 7156 digits. It has 7149 consecutive zeros between the digits 217833 and 1.

168 prime numbers between 1 and 1000
135 prime numbers between 1000 and 2000
127 prime numbers between 2000 and 3000
120 prime numbers between 3000 and 4000
119 prime numbers between 4000 and 5000
114 prime numbers between 5000 and 6000
117 prime numbers between 6000 and 7000
107 prime numbers between 7000 and 8000
110 prime numbers between 8000 and 9000
112 prime numbers between 9000 and 10000
106 prime numbers between 10000 and 11000
103 prime numbers between 11000 and 12000
109 prime numbers between 12000 and 13000
105 prime numbers between 13000 and 14000
102 prime numbers between 14000 and 15000
108 prime numbers between 15000 and 16000
98 prime numbers between 16000 and 17000
104 prime numbers between 17000 and 18000
94 prime numbers between 18000 and 19000
104 prime numbers between 19000 and 20000
98 prime numbers between 20000 and 21000
104 prime numbers between 21000 and 22000
100 prime numbers between 22000 and 23000
104 prime numbers between 23000 and 24000
94 prime numbers between 24000 and 25000
98 prime numbers between 25000 and 26000
101 prime numbers between 26000 and 27000
94 prime numbers between 27000 and 28000
98 prime numbers between 28000 and 29000
92 prime numbers between 29000 and 30000
95 prime numbers between 30000 and 31000
92 prime numbers between 31000 and 32000
106 prime numbers between 32000 and 33000
100 prime numbers between 33000 and 34000
94 prime numbers between 34000 and 35000
92 prime numbers between 35000 and 36000
99 prime numbers between 36000 and 37000
94 prime numbers between 37000 and 38000
90 prime numbers between 38000 and 39000
96 prime numbers between 39000 and 40000
88 prime numbers between 40000 and 41000
101 prime numbers between 41000 and 42000
102 prime numbers between 42000 and 43000
85 prime numbers between 43000 and 44000
96 prime numbers between 44000 and 45000
86 prime numbers between 45000 and 46000
90 prime numbers between 46000 and 47000
95 prime numbers between 47000 and 48000
89 prime numbers between 48000 and 49000
98 prime numbers between 49000 and 50000

Figure 2-1
Prime numbers become scarce when computed in large number ranges.

NUMBER THEORY TRIVIA

Prime numbers which read the same both forward and backward are called **palindromes**. Examples are 101, 131, 313, 383, 727, 757, 787, 797, and 919.

For large numbers this method may be very cumbersome; however, here as in many other calculations of number theory, you can rely on modern computational techniques. It is simple to program a computer to divide a given number by all integers up to the square root of the number and to print those which give no remainder.

The following program uses this technique to produce all primes less than 10,000. Since we know immediately and automatically that 2 is the only even prime, the program examines only the odd numbers starting with the number 3, and divides each succeeding odd number up to the square root of 10,000 by all primes that are found. Figure 2-2 lists the first 506 primes that were produced by this program.

```
program PrimeNumbers;
var
  I,J,K,N       :Integer;
begin
  writeln('LIST OF PRIME NUMBERS');
  writeln('------------------------------');
  writeln;
  K := 1;
  while K < 5000 do
    begin
      N := 2 * K + 1;
      J := 1;
      I := 3;
      while (I * I <= N) and (J = 1) do
        if N mod I = 0
          then J := 0
          else I := I + 2;
      if J = 1 then writeln(N);
      K := K + 1;
    end;
end.
```

NUMBER THEORY TRIVIA

Prime triplets are a set of prime numbers that are determined by one of the following sequences:

n n+2 n+4
such as 3, 5, 7

n n+2 n+6
such as 5, 7, 11 or 11, 13, 17

n n+4 n+6
such as 7, 11, 13 or 13, 17, 19

Can you write a program to generate several prime triplets?

LIST OF PRIME NUMBERS

3	5	7	11	13	17	19	23	29	31	37
41	43	47	53	59	61	67	71	73	79	83
89	97	101	103	107	109	113	127	131	137	139
149	151	157	163	167	173	179	181	191	193	197
199	211	223	227	229	233	239	241	251	257	263
269	271	277	281	283	293	307	311	313	317	331
337	347	349	353	359	367	373	379	383	389	397
401	409	419	421	431	433	439	443	449	457	461
463	467	479	487	491	499	503	509	521	523	541
547	557	563	569	571	577	587	593	599	601	607
613	617	619	631	641	643	647	653	659	661	673
677	683	691	701	709	719	727	733	739	743	751
757	761	769	773	787	797	809	811	821	823	827
829	839	853	857	859	863	877	881	883	887	907
911	919	929	937	941	947	953	967	971	977	983
991	997	1009	1013	1019	1021	1031	1033	1039	1049	1051
1061	1063	1069	1087	1091	1093	1097	1103	1109	1117	1123
1129	1151	1153	1163	1171	1181	1187	1193	1201	1213	1217
1223	1229	1231	1237	1249	1259	1277	1279	1283	1289	1291
1297	1301	1303	1307	1319	1321	1327	1361	1367	1373	1381
1399	1409	1423	1427	1429	1433	1439	1447	1451	1453	1459
1471	1481	1483	1487	1489	1493	1499	1511	1523	1531	1543
1549	1553	1559	1567	1571	1579	1583	1597	1601	1607	1609
1613	1619	1621	1627	1637	1657	1663	1667	1669	1693	1697
1699	1709	1721	1723	1733	1741	1747	1753	1759	1777	1783
1787	1789	1801	1811	1823	1831	1847	1861	1867	1871	1873
1877	1879	1889	1901	1907	1913	1931	1933	1949	1951	1973
1979	1987	1993	1997	1999	2003	2011	2017	2027	2029	2039
2053	2063	2069	2081	2083	2087	2089	2099	2111	2113	2129
2131	2137	2141	2143	2153	2161	2179	2203	2207	2213	2221
2237	2239	2243	2251	2267	2269	2273	2281	2287	2293	2297
2309	2311	2333	2339	2341	2347	2351	2357	2371	2377	2381
2383	2389	2393	2399	2411	2417	2423	2437	2441	2447	2459
2467	2473	2477	2503	2521	2531	2539	2543	2549	2551	2557
2579	2591	2593	2609	2617	2621	2633	2647	2657	2659	2663
2671	2677	2683	2687	2689	2693	2699	2707	2711	2713	2719
2729	2731	2741	2749	2753	2767	2777	2789	2791	2797	2801
2803	2819	2833	2837	2843	2851	2857	2861	2879	2887	2897
2903	2909	2917	2927	2939	2953	2957	2963	2969	2971	2999
3001	3011	3019	3023	3037	3041	3049	3061	3067	3079	3083
3089	3109	3119	3121	3137	3163	3167	3169	3181	3187	3191
3202	3209	3217	3221	3229	3251	3253	3257	3259	3271	3299
3301	3307	3313	3319	3323	3329	3331	3343	3347	3359	3361
3371	3373	3389	3391	3407	3413	3433	3449	3457	3461	3463
3467	3469	3491	3499	3511	3517	3527	3529	3533	3539	3541
3547	3557	3559	3571	3581	3583	3593	3607	3613	3617	3623

Figure 2-2
The first 506 odd primes produced by the Prime Number program. The output is shown in a condensed form to conserve space. The program would actually print the prime numbers in a single column.

NUMBER THEORY TRIVIA

A **palindromic prime** is a prime number that reads the same forward and backward. For example 11, 131, 353, or 383. The following table lists some of the facts known about palindromic primes.

Number of Decimal Digits	Total Number of Primes	Number of Palindromic Primes
1	4	4
2	21	1
3	143	15
4	1061	0
5	8363	93
6	68906	0
7	586081	668
8	5096876	0

NUMBER THEORY TRIVIA

In 1964, Paul Levy, a French mathematician, postulated that every odd number greater than 3 can be expressed in the form 2P + Q where P and Q are prime:

21 = 2(5) + 11
27 + 2(11) + 5
35 = 2(11) + 13
77 = 2(37) + 3
93 = 2(43) + 7
121 = 2(59) + 3
169 = 2(43) + 83
437 = 2(197) + 43
445 = 2(211) + 23
999 = 2(283) + 433
1057 = 2(487) + 83

Levy's conjecture has been neither verified nor disproven

Now look at a program that will determine whether a given positive integer (N) is a prime number. In this program, one types the number on a keyboard. The program determines whether N is prime and so indicates in an output message. To determine if the input number is prime this program attempts to divide that number by all possible integers betwen 2 and $\sqrt{N}$. Program and sample results are as follows.

```
program PrimeCheck;
var
  N      :Integer;(* Number to check *)
  I      :Integer;(* Looping counter *)
begin
  (**** Input an integer to check ****)
  write('ENTER AN INTEGER GREATER THAN 1: ');
  readln(N);
    for I := 2 to TRUNC(SQRT(N)) do
      if N div I = N/I then
        begin
          writeln(N,' IS NOT A PRIME');
          HALT;
        end;
    writeln(N,' IS A PRIME NUMBER');
end.
```

COMPILE/RUN

ENTER AN INTEGER GREATER THAN 1: 1399
1399 IS A PRIME NUMBER

ENTER AN INTEGER GREATER THAN 1: 227
227 IS A PRIME NUMBER

ENTER AN INTEGER GREATER THAN 1: 1251
1251 IS NOT A PRIME

ENTER AN INTEGER GREATER THAN 1: 3511
3511 IS A PRIME NUMBER

NUMBER THEORY TRIVIA

There is a set of prime numbers consisting of digits in consecutive order. The following table shows primes with digits in ascending order (01234567890123. . .).

23
67
89
4567
67891
78901
89123
678901
4567891
23456789
45678901
56789123
1234567891
4567891234567
9012345678901
789012345678901
890123456789123
1234567890123456789I

Can you write a program to find other prime numbers in ascending order? In descending order?

The trial-division algorithm used in the previous programs is incapable of testing the largest primes known. Consider the 13,395 digit prime number $2^{44,497} - 1$. If a computer were to carry out trial divisions at the rate of a million divisions per second, and if it were to stop once it reached the square root of the number, it would need $10^{6,684}$ years to finish the task. The trouble with the trial-division method is that it does far more than is required: trial-division not only decides whether a number is prime or composite but also gives factors of any composite number. Although there are methods of factoring that do not depend on trial-division, none of them can factor an arbitrary number having a "mere" 100 digits in any reasonable time, even with a large computer. It turns out, however, that it is possible to determine whether or not a number is prime without necessarily finding any factors in case the number is composite. If the number has no small factors, such methods are almost invariably more efficient than the methods that give the factors. Mathematicians have developed a method that enables a computer to determine the primality of an arbitrary 100-digit number in about 40 seconds of running time.

Just think — while I'm lying here, my computer is trying to find the next largest prime number.

NUMBER THEORY TRIVIA

A **repunit prime** is a prime number consisting only of two or more 1's. The smallest repunit prime is 11. Repunits are usually abbreviated R_n, where *n* represents the number of 1's to be written. 11 is R_2. Other repunit primes are:

R_{19} = 1111 11111 11111 11111
R_{23} = 111 11111 11111 11111 11111

2.2 THE SIEVE OF ERATOSTHENES

One method for finding primes smaller than some given number was first used by the ancient Greek scholar Eratosthenes more than 2000 years ago. This technique is known as the Sieve of Eratosthenes. Suppose we wish to find the primes less than 100. First, we prepare a table of the numbers through 100, as shown in Figure 2-3.

Figure 2-3
Finding primes using the Sieve of Eratosthenes.

~~1~~	**2**	**3**	~~4~~	**5**	~~6~~	**7**	~~8~~	~~9~~	~~10~~
11	~~12~~	**13**	~~14~~	~~15~~	~~16~~	**17**	~~18~~	**19**	~~20~~
~~21~~	~~22~~	**23**	~~24~~	~~25~~	~~26~~	~~27~~	~~28~~	**29**	~~30~~
31	~~32~~	~~33~~	~~34~~	~~35~~	~~36~~	**37**	~~38~~	~~39~~	~~40~~
41	~~42~~	**43**	~~44~~	~~45~~	~~46~~	**47**	~~48~~	~~49~~	~~50~~
~~51~~	~~52~~	**53**	~~54~~	~~55~~	~~56~~	~~57~~	~~58~~	**59**	~~60~~
61	~~62~~	~~63~~	~~64~~	~~65~~	~~66~~	**67**	~~68~~	~~69~~	~~70~~
71	~~72~~	**73**	~~74~~	~~75~~	~~76~~	~~77~~	~~78~~	**79**	~~80~~
~~81~~	~~82~~	**83**	~~84~~	~~85~~	~~86~~	~~87~~	~~88~~	**89**	~~90~~
~~91~~	~~92~~	~~93~~	~~94~~	~~95~~	~~96~~	**97**	~~98~~	~~99~~	~~100~~

Cross out 1, since it is not classified as a prime number.

Draw a circle around 2, the smallest prime number. Then cross out every following multiple of 2, since each one is divisible by 2 and thus is not prime.

Draw a circle around 3, the next prime number. Then cross out each succeeding multiple of 3. Some of these numbers, such as 6 and 12, will already have been crossed out because they are also multiples of 2.

Circle the next open number, 5, and cross out all subsequent multiples of 5.

The next open number is 7; circle 7 and cross out multiples of 7. Since 7 is the largest prime less than $\sqrt{100}$, it follows that all of the remaining numbers are prime.

Eratosthenes (275-194 B.C.), a Greek geographer and mathematician, was one of the greatest scholars of Alexandria. He was a highly talented and versatile person. In addition to his Sieve for prime numbers, he made some remarkably accurate measurements of the size of the earth.

The process is a simple one, since you do not have to cross out the multiples of 3 (for example) by checking for divisibility by 3 but can simply cross out every third number. Thus anyone who can count can find primes by this method. Also, notice that, in finding the primes under 100, we had crossed out all the composite numbers by the time we crossed out the multiples of 7. That is, to find all primes less than 100: (1) Find the largest prime smaller than or equal to the square root of 100 (7 in this case); (2) cross out multiples of primes up to and including 7; and (3) all the remaining numbers in the chart are primes.

The following program computes all primes up to 1000 using the Sieve of Eratosthenes.

```
program SieveOfEratosthenes;
var
  N       :array[1..1000] of Integer;(* Number array *)
  P       :array[1..200] of Integer;(* Prime number array *)
  I       :Integer;(* Loop variable *)
  K,R     :Integer;(* Counters *)
begin
  for I := 2 to 1000 do
    N[I] := 0;
  (**** Determine prime numbers ****)
  K := 0;
  for R := 2 to 1000 do
    if N[R] = 0 then
      begin
        K := K + 1;
        P[K] := R;
        if R <= INT(SQRT(1000)) then
          begin
            I := R;
            repeat
              N[I] := -1;
              I := I + R;
            until I > 1000;
          end;
      end;
  (**** Print prime numbers ****)
  for I := 1 to K do
    write(P[I]:3, ' ');
end.
```

A microcomputer program can find primes using the Sieve of Eratosthenes.

The program uses two arrays. Array N is to be thought of as the list of numbers from 2 to 1000, with N(I) standing for I. The program crosses out the number I by setting N(I) = -1. Array P contains the primes. The number K indicates how many primes we have found.

NUMBER THEORY TRIVIA

The following four-digit prime numbers are three permutations of the same digit sets:

1423	2341	2143
1847	8147	8741
1933	3391	3931
2617	2671	6271
2633	3623	6323
3539	3593	5393
3797	7937	9377
4583	5483	8543
4657	5647	6547
8699	8969	9689

Can you write a program to find additional four-digit primes which are three permutations of the same digit sets?

The program is initialized by storing zeros in array N and setting K to 0. Then the program performs the sieve of Eratosthenes. The program picks the next number P in array N. If it has been crossed out, it is forgotten. If not, it is added to array P and all of its multiples are crossed out. Finally the program prints the list of prime numbers.

2.3 WILSON'S THEOREM

Take all the integers from 1 up to some prime number, say 11, and write their product 1 × 2 × 3 × 4 × 5 × 6 × 7 × 8 × 9 × 10 × 11. Symbolically this is condensed into 11! and called "eleven factorial." It is easy to see that 11! is exactly divisible by every integer from 1 up to and including 11. Now omit the 11, and it is clear that the product of the remaining 10 integers, namely, 10! is *not* divisible by 11, since 11 is a prime, and a product not containing 11 or one of its multiples can obviously never be divided by 11. But if we add 1 to the product - presto! the result *is* divisible by 11. Similarly 6! + 1 = (1 × 2 × 3 × 4 × 5 × 6) + 1 = 721 is divisible by 7. This holds true in the above two examples because 7 and 11 are prime numbers and it is ***always*** true for prime numbers and *never* true for composite numbers.

Generalizing the above, (p − 1)! + 1 is divisible by p if, and only if, p is a prime. This is called Wilson's theorem and some of its corollaries and generalizations are very important.

Theoretically, the theorem is a criterion for determining whether or not a number is a prime. So, to determine whether 8191 is a prime, it would be necessary only to multiply together all the numbers from 1 to 8190, add 1 and see whether 8191 divides this number exactly, in which case 8191 is a prime; if not, it is composite. Thus, in theory, the problem of determining whether a given number is prime is completely solved. But for large integers, the computational difficulties are great.

The program on page 45 uses Wilson's theorem to compute a table of values for the integers 2 through 8. The reader can increase the size of this table by modifying the **for** statement. Remember, however, that (N − 1)! will become a large number very quickly as shown below:

N	(N − 1)!
12	39 916 800
13	479 001 600
14	6 227 020 800
15	87 178 291 200

Many microcomputers cannot perform computations with integers of this size.

NUMBER THEORY TRIVIA

Wilson's theorem states that $(p - 1)! + 1$ is always exactly divisible by p when, and only when, p is a prime number. This same expression is, on very rare occasions, divisible not only by p but by p^2 as well. The only primes less than 100,000 for whch this is true are 5, 13, and 563.

```
program WilsonsTheorem;
var
  Number      :Integer;(* Number *)
  C1          :Integer;(* (N - 1)Factorial *)
  C2          :Integer;(* (N - 1) Factorial + 1 *)
  C3          :Integer;(* ((N - 1) Factorial + 1) / N *)
function Factorial (N:Integer):Real;
begin (**** Factorial function ****)
  if N = 0
    then Factorial := 1
    else Factorial := N * Factorial(N - 1);
end; (**** End of Factorial function ****)
begin
  writeln('N     (N-1)!     (N-1)!+1     REMAINDER OF     CHARACTER');
  writeln('                              ((N-1)!+1)/N');
  writeln('-------------------------------------------------------------');
  for Number := 2 to 8 do
    begin
      (**** Apply Wilsons theorem ****)
      C1 := TRUNC(Factorial(Number - 1));
      C2 := C1 + 1;
      C3 := C2 mod Number;
      (**** Print table ****)
      write(Number,C1:12,C2:22,C3:17);
      if C3 <> 0
        then writeln('COMPOSITE':20)
        else writeln('PRIME':20);
    end;
end.
```

COMPILE/RUN

N	(N−1)!	(N−1)!+1	REMAINDER OF ((N−1)!+1)/N	CHARACTER
2	1	2	0	PRIME
3	2	3	0	PRIME
4	6	7	3	COMPOSITE
5	24	25	0	PRIME
6	120	121	1	COMPOSITE
7	720	721	0	PRIME
8	5040	5041	1	COMPOSITE

NUMBER THEORY TRIVIA

The prime 1193 is reversible to 3911. The number 3911 is also a prime. Can you write a program to produce all 4-digit **reversible primes**? A few of them are 1009, 1249, 1471, 1657, 1741, 1789, 1867, 3257, 3391, 3697, 3911, 7561, 7681, 7963, 9001, 9349, 9421, 9439, 9661, and 9871.

2.4 RARE PRIME DECADES

Assume that a decade is a set of ten consecutive numbers. The first decade is from 1 to 10 inclusive, the second decade from 11 to 20 and so on. In each such decade there can never be more than four prime numbers. In the vast percentage of cases the number of primes in a decade is two or three and four primes are quite rare. For example from 51 to 60 we have 53 and 59 as the only primes. From 71 to 80 we have 71, 73 and 79. Very seldom indeed do four primes appear in a decade.

The rarity of four primes in a decade is due to the fact that there are only four possible numbers in a decade that can be prime. In each decade there are 5 even numbers which, with the exception of 2, can never be prime, and five odd numbers. Of the five odd numbers there is always a multiple of 5 such as 35, 55, 85, etc., and, of course, this cannot be prime. The remaining four numbers must end in 1, 3, 7 and 9, no matter how large they are, and often one or another of these is divisible by 3 or 7 or 9, such as, for example, 21, 49, 63, etc.

The following program computes a list of rare primes up to 5000.

```
program RarePrimeDecades;
var
  N      :Integer;(* Prime number *)
  K      :Integer;(* Counter *)
  I,J    :Integer;(* Computational variable *)
  D      :Integer;(* Decade counter *)
  C      :Integer;(* Primes within decade counter *)
  P      :array[1..4] of Integer;
begin
  (**** Initialize counters ****)
  D := 10;
  C := 1;
  writeln('PRIME DECADES');
  writeln('-------------------');
  (**** Set even prime number ****)
  P[C] := 2;
  C := C + 1;
  (**** Compute prime numbers ****)
  K := 1;
  while K < 2500 do
    begin
      N := 2 * K + 1;
      J := 1;
      I := 3;
      while(I * I <= N) and (J = 1) do
        if N mod I = 0
          then J := 0
          else I := I + 2;
```

NUMBER THEORY TRIVIA

The reverse of the prime 61 is 16 whose square root is 4. Other primes of this form are:

PRIME	REVERSE PRIME	SQUARE ROOT
163	361	19
487	784	28
691	196	14
1297	7921	89
1861	1681	41
4201	1024	32
4441	1444	38
4483	3844	62
5209	9025	95
5227	7225	85
9049	9409	97

Can you write a program to produce additional primes of this form?

```
        if J = 1 then
          begin
            if N <= D then
              begin
                P[C] := N;
                C := C + 1;
              end
              else
                begin
                  (**** Are there 4 primes in decade? ****)
                  if C = 5 then
                    begin
                      C := 1;
                      (**** Print primes in decade ****)
                      writeln(P[C]:4, ' ', P[C+1]:8, ' ', P[C+2]:8, ' ', P[C+3]:8);
                    end;
                  (**** Reset prime counter ****)
                  C := 1;
                  P[C] := N;
                  C := C + 1;
                  (**** Increment decade counter ****)
                  repeat
                    D := D + 10;
                  until N <= D;
                end;
          end;
        K := K + 1;
      end;
end.
```

2.5 MERSENNE NUMBERS

A prime race has been going on for several centuries. Many mathematicians and computer scientists have vied for the honor of having discovered the greatest known prime. Historically, the largest known prime has been a Mersenne Prime.

Mersenne Numbers are numbers that can be expressed as $2^p - 1$, where p is a prime number, and are denoted M_p. For example $M_{11} = 2^{11} - 1 = 2047$. Mersenne Numbers are named after a French monk, **Father Marin Mersenne** (1588-1648). Mersenne was an amateur mathematician who corresponded regularly with Fermat and other mathematicians of his day.

The first few Mersenne primes are

$M_2 = 2^2 - 1 = 3$ $\qquad$ $M_{13} = 8191$

$M_3 = 2^3 - 1 = 7$ $\qquad$ $M_{17} = 131{,}071$

$M_5 = 2^5 - 1 = 31$ $\qquad$ $M_{19} = 524{,}287$

$M_7 = 127$ $\qquad$ $M_{31} = 2{,}147{,}483{,}647$

```
PRIME DECADES
-------------------
2       3       5       7
11      13      17      19
101     103     107     109
191     193     197     199
821     823     827     829
1481    1483    1487    1489
1871    1873    1877    1879
2081    2083    2087    2089
3251    3253    3257    3259
3461    3463    3467    3469
```

Output from Rare Prime Decades program.

Marin Mersenne (1588-1648) was a Franciscan friar and spent most of his life in Parisian cloisters. His importance to the mathematical world of his day was due not so much to his own work, as to his energy, friendly enthusiasm, and extensive correspondence with other mathematicians.

NUMBER THEORY TRIVIA

It is not known whether or not there is an infinite number of Mersenne primes.

NUMBER THEORY TRIVIA

Look at the prime number

73,939,133

By dropping successively the right-most digit, the remaining numbers are still all primes. The prime numbers which can be generated in this manner are

73,939,133
7,393,913
739,391
73,939
7,393
739
73
7

These numbers are sometimes called **prime primes**.

The following program will produce a Mersenne Number for a specified input prime.

```
program MersenneNumbers;
var
  P       :Integer;(* Prime number *)
  M       :Integer;(* Mersenne number *)
function POWER(X:Real;K:Integer):Real;
(**** This function returns the Kth power of X ****)
var
  Result     :Real;
  Index      :Integer;
begin
  Result := 1;
  for Index := 1 to K do
    Result := Result * X;
  POWER := Result;
end; (* FUNCTION POWER *)
begin
  write('ENTER PRIME NUMBER: ');
  readln(P);
  (**** Determine Mersenne number ****)
  M := TRUNC(POWER(2.0,P) - 1);
  (**** Print Mersenne number ****)
  writeln('MERSENNE NUMBER - ',M);
end.
```

The general procedure for finding large primes of the Mersenne type is to examine all the numbers M_p for the various primes p. The numbers increase very rapidly and so do the labors involved. The reason why the work is manageable even for quite large numbers is that there are very effective ways for determining whether these special numbers are primes.

There was an early phase in the examination of Mersenne primes which culminated in 1750 when the Swiss mathematician, Leonhard

Mersenne prime #	M_p where p =
1st	2
2nd	3
3rd	5
4th	7
5th	13
6th	17
7th	19
8th	31
9th	61
10th	89
11th	107
12th	127
13th	521
14th	607
15th	1279
16th	2203
17th	2281
18th	3217
19th	4253
20th	4423
21st	9689
22nd	9941
23rd	11213
24th	19937
25th	21701
26th	23209
27th	44497
28th	86243
29th	132049
30th	216091
31st	110503

The known Mersenne primes. The 31st prime occurs in a gap between two previous-known Mersenne primes.

Euler, established that M_{31} is a prime. By that time only eight Mersenne primes had been found. In 1876, the French mathematician Lucas established that the 39-digit number

$$M_{127} = 170\ 141\ 183\ 460\ 469\ 231\ 731\ 687\ 303\ 715\ 884\ 105\ 727$$

is a prime. The introduction of electronic hand calculators made it possible to find Mersenne primes up to M_{257}, but the results were disappointing; no further Mersenne primes were found.

In 1876, Edouard Lucas discovered a fast way to test the primality of a Mersenne Number. Using the test and calculators, several primes were added to the list of Mersenne Primes. In 1930, D. H. Lehmer published an improved version of Lucas' algorithm. The Lucas-Lehmer test for primality is:

Let $u(1) = 4$.
For $i = 1$ to $p - 2$ compute $u(i + 1) = (u(i)^2 - 2) \bmod M_p$.
If and only if $u(p - 1) = 0$, then M_p is a prime.

The "mod M_p" means to keep only the remainder after division by M_p.

By way of example, let us apply the Lucas-Lehmer test to $M_5 = 31$.

$$u(1) = 4$$
$$u(2) = (4^2 - 2) \bmod 31 = 14$$
$$u(3) = (14^2 - 2) \bmod 31 = 8$$
$$u(4) = (8^2 - 2) \bmod 31 = 0$$

Hence, M_5 is indeed a prime number.

This was the situation when computers came into being. With the development of newer and larger computers it was possible to push the search for Mersenne primes further and further. By 1963, Mersenne primes had been computed up to M_{11213}, a prime number containing 3376 digits. This prime was determined using number theory techniques and a computer at the University of Illinois. The mathematicians were so proud of this discovery that they advertised it on the university's postage meter.

$2^{11213} - 1$
IS PRIME

The twenty-fourth Mersenne prime has 6002 digits. In 1971, Bryant Tuckerman using an IBM System 360/91 computer found M_{19937} in 39.44 minutes.

As bigger and faster computers were built and better programs developed, new Mersenne primes turned up. In 1978, two high school students, Curt Noll and Laura Nickel, discovered the 25th

LARGEST
MERSENNE PRIME

- M_{216091}
- 65,050 DIGITS LONG
- DISCOVERED BY DAVID SLOWINSKI

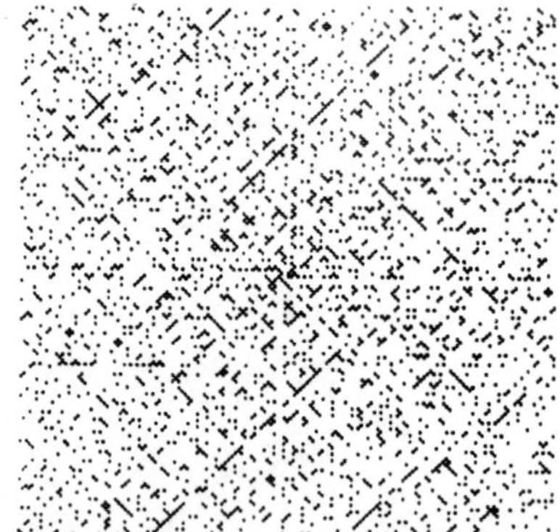

Stanislaw Ulam, the Polish-American mathematician, discovered an interesting graphical image for primes (shown above). After plotting the primes in a counter-clockwise spiraling direction, he noticed an odd pattern developing. Straight lines began to form out of nowhere! The prominent diagonal lines correspond to prime lodes. Near the center of the diagram one such deposit proceeds down and to the left. It consists of the number sequence 7, 23, 47, 79, . . . The formula for this prime number sequence is the quadratic $4x^2 + 4x - 1$. Other similar formulas can be written for virtually any line in the diagram.

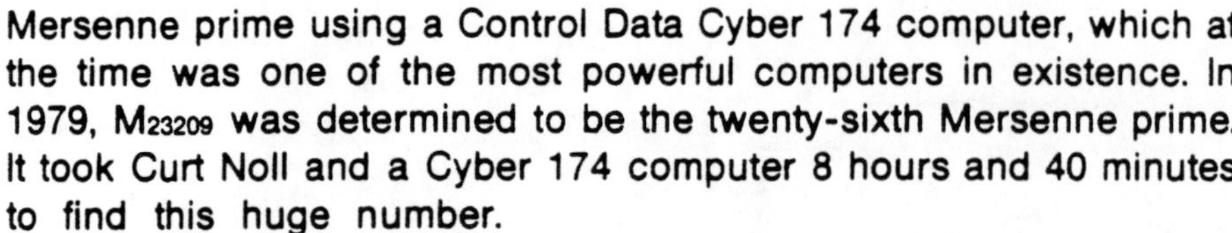

Mersenne prime using a Control Data Cyber 174 computer, which at the time was one of the most powerful computers in existence. In 1979, M_{23209} was determined to be the twenty-sixth Mersenne prime. It took Curt Noll and a Cyber 174 computer 8 hours and 40 minutes to find this huge number.

In April, 1979, the 27th Mersenne prime, M_{44497}, was found during the acceptance testing of a Cray-1 supercomputer at the Lawrence Livermore Laboratories. The 27th prime is a 13,395 digit whopper, but it's small potatoes next to its big brother, Mersenne prime 28. In 1982, a Cray supercomputer was used by David Slowinski of Cray Research Corporation to find this number. Working at speeds of 80 million multiplications and 80 million additions every second, the Cray-1 supercomputer found the 28th Mersenne prime ($2^{86243} - 1$) after some 600 hours of computer time.

The 29th Mersenne prime was found by David Slowinski using a Cray X-MP supercomputer. The Cray X-MP is capable of doing more than 800 million multiplications and divisions each second. In verifying the primality of the 29th Mersenne prime, it took the computer slightly more than one hour.

In the continuing search for larger prime numbers, David Slowinski, a computer scientist at Cray Research in Chippewa Falls, Wisconsin, discovered the 30th Mersenne prime during Labor Day weekend in 1985 while testing a newly installed Cray X-MP 24 supercomputer at the Chevron Geosciences facility in Houston, Texas. This new prime number contains 65,050 digits and fills 14 pages of computer paper. It took three hours to verify the primality of the 30th Mersenne prime.

Because the testing for the last few Mersenne primes is not being done systematically, there is no guarantee that all the numbers between M_{86243} and M_{216091} have been checked. It is possible that there are other Mersenne primes in between.

In 1988, Walter N. Colquitt (Houston Area Research Center) and Luther Welsh, Jr., using an NEC FX-2 supercomputer, found the 31st Mersenne prime:

$$2^{110503} - 1$$

This Mersenne prime occurs in a gap between two previous-known Mersenne primes.

Are there an infinite number of Mersenne primes? No one knows the answer to this question.

David Slowinski, a computer scientist at Cray Research Corporation, has used Cray supercomputers for generating several large Mersenne primes: M_{44497}, M_{86243}, M_{132049}, and M_{216091}.

Pierre de Fermat (1601-1665) was not a mathematician by profession, but a government official. He did not interest himself in mathematics until he was past thirty. Fermat is best known for his work in number theory, notably on primes.

2.6 FERMAT NUMBERS

A family of numbers related to the Mersenne numbers are those of the form

$$F_t = 2^{2^t} + 1$$

These numbers are known as the Fermat numbers, and for the first values of t they seem to be primes.

$$F_0 = 3$$
$$F_1 = 5$$
$$F_2 = 17$$
$$F_3 = 257$$
$$F_4 = 65{,}537$$

The next Fermat number is already so large that it is difficult to factor, but on the basis of the few facts at hand Fermat made the conjecture that they are all primes. It was not until 100 years later (1739) that Leonhard Euler exploded the hypothesis by showing that the next Fermat number had a factor. Fermat numbers have been the object of numerous studies, both theoretical and computational, and quite a few of the larger Fermat numbers have been successfully factored. Of Fermat's original conjecture there is no trace; no further primes have been found. Students of the question now seem more inclined to the opposite hypothesis that there are no further Fermat primes than the first five already found.

NUMBER THEORY TRIVIA

In 1987, Jeff Young and Duncan A. Buell, using a Cray-2 supercomputer proved that

$$F_{20} = 2^{2^{20}} + 1$$

is composite. For years, F_{20} had been the smallest Fermat number of unknown character. F_{22} is now the smallest Fermat number of unknown character.

2.7 TWIN PRIMES

Two consecutive odd integers that are prime are called twin primes. The first few pairs of twin primes are 3 and 5, 5 and 7, 11 and 13, 17 and 19, 29 and 31, 41 and 43. The following program finds and prints all pairs of twin primes less than 1000.

ISN'T IT EXCITING TO BE TWIN PRIMES?

```
program TwinPrimes;
var
  Prime                       :array[1..400] of Integer;
  Ctr,Num,A,B,PrimeIndex      :Integer;
begin
  PrimeIndex := 1;
  (**** Determine prime numbers less than 1000 ****)
  Ctr := 1;
  while Ctr < 500 do
    begin
      Num := 2 * Ctr + 1;
      A := 1;
      B := 3;
      while (B * B <= Num) and (A = 1) do
        if Num mod B = 0
          then A := 0
          else B := B + 2;
```

NUMBER THEORY TRIVIA

A large twin prime pair is

$$76 \times 3^{139} \pm 1$$

or

158,733,282,881,841,916,274,491, 012,923,328,901,749,236,259,319, 203,520,296,443,150,620,292 ± 1

```
        if A = 1 then
          begin
            Prime[PrimeIndex] := Num;
            PrimeIndex := PrimeIndex + 1;
          end;
        Ctr := Ctr + 1;
end;
  (**** Determine and print twin primes ****)
  writeln('TWIN PRIMES');
  writeln('----------------');
  for PrimeIndex := 1 to 170 do
    if Prime[PrimeIndex + 1] - Prime[PrimeIndex] = 2
      then writeln(Prime[PrimeIndex],'     ',Prime[PrimeIndex + 1]);
end.
```

```
TWIN PRIMES
----------------
3        5
5        7
11       13
17       19
29       31
41       43
59       61
71       73
101      103
107      109
137      139
149      151
179      181
191      193
197      199
227      229
239      241
269      271
281      283
311      313
347      349
419      421
431      433
461      463
521      523
569      571
599      601
617      619
641      643
659      661
809      811
821      823
827      829
857      859
881      883
```

NUMBER THEORY TRIVIA

There are several rare **nine-digit twin primes** in which each member of the pair forms a reversible prime. For example,

173313197	—	791313371
173313199	—	991313371

and

791973311	—	113379197
791973313	—	313379197

The primes contain only the digits 1, 3, 7, and 9.

Can you write a program to find all 49 pairs of nine-digit twin primes that have this property?

NUMBER THEORY TRIVIA

Prime numbers differing by 2 are called twin primes. Examples are 11 and 13, and 17 and 19. A prime number will be considered **isolated** provided it is not an element of a set of twin primes. The following table shows the isolated primes less than 500.

23	113	211	307	401
37	127	223	317	409
47	131	233	331	439
53	157	251	337	443
67	163	257	353	449
79	167	263	359	457
83	173	277	367	467
89		293	373	479
97			379	487
			383	491
			389	499
			397	

It has been proven that there are infinitely many prime numbers, however, we still do not know whether there are an infinite number of twin primes. Consequently, mathematicians have been looking for large twin primes for centuries.

2.8 PRIME POLYNOMIALS

For centuries, mathematicians have tried to find a formula that would yield every prime - or even a formula that would yield only primes. No one has yet found such a formula, but several remarkable expressions produce large numbers of primes for consecutive values of x. For example, $2x^2 + 29$ will give primes (starting with 29) for x = 0 to 28 (twenty-nine primes), $x^2 + x + 41$ will give primes for x = 0 to 39 (forty primes starting with 41); $x^2 + x + 17$ will generate sixteen primes, $6x^2 + 6x + 31$ will give primes for twenty-nine values of x, $3x^2 + 3x + 23$ will give primes for twenty-two values of x, and $x^2 - 79x + 1601$ will give eighty consecutive prime values when x = 0, 1, 2, . . ., 79. Other examples of the same nature exist.

The following program uses the formula

$$x^2 - x + 41$$

to generate primes for the 40 values of x: 1, 2, 3, . . ., 40.

```
program Prime Polynomial;
var
  X,P      :Integer;
begin
  writeln('PRIME  NUMBERS');
  for X := 1 to 40 do
    begin
      P := X * X - X + 41;
      writeln(P);
    end;
end.
```

COMPILE/RUN

PRIME NUMBERS

41	43	47	53	61
71	83	97	113	131
151	173	197	223	251
281	313	347	383	421
461	503	547	593	641
691	743	797	853	911
971	1033	1097	1163	1231
1301	1373	1447	1523	1601

NUMBER THEORY TRIVIA

There are two sets of four-digit twin primes of the type $10^n a + a \pm 1$, wherein a consists of n digits. One set of twin primes is

4241 and 4243

Can you write a program to find the other set of four-digit twin primes?

2.9 EMIRP'S

An emirp (prime spelled backwards) is a prime number whose reversal is a different prime number. For example, 17 is an emirp because 71 is also a prime. Other examples of emirps are 107 (and 701) and 113 (and 311). Perhaps the reader would be interested in writing a program to find all emirps less than 5000.

NUMBER THEORY TRIVIA

Emirps must occur in pairs. The only two-digit emirps are 13-31, 17-71, 37-73, and 79-97. However, there are 13 pairs of three-digit emirps, 102 pairs of four-digit emirps, and at least 684 pairs of five-digit emirps.

NUMBER THEORY TRIVIA

No one knows whether **emirps** go on forever or whether they end with a large emirp.

REVIEW EXERCISES

1. What is a prime number?
2. Who first proved that there are infinitely many prime numbers?
3. Is every odd number a prime number?
4. Is every number a composite number?
5. Show that 14 is the sum of two primes.
6. Give an example of five consecutive composite numbers less than 60.
7. What is the only even prime number?
8. Which of the following numbers are primes?
   (a) The year of your birth?
   (b) The present year number?
   (c) Your house number?
   (d) Your girlfriend's/boyfriend's age?
9. There are
   25 prime numbers between 1 and 100
   21 prime numbers between 100 and 200
   16 prime numbers between 200 and 300
   16 prime numbers between 300 and 400
   17 prime numbers between 400 and 500
   14 prime numbers between 500 and 600
   16 prime numbers between 600 and 700
   14 prime numbers between 700 and 800
   15 prime numbers between 800 and 900
   14 prime numbers between 900 and 1000
   Write a program that will continue this table of prime numbers up to 3000.
10. Identify a pair of prime numbers that differ by 1. Why are they the only such pair?
11. How many primes are there between 1 and 5000?
12. Which two consecutive primes less than 1000 have the greatest difference between them?
13. Modify the Prime Numbers program so that it will generate 20,000 prime numbers.
14. Write a program that will generate primes between 1000 and 2000.

NUMBER THEORY TRIVIA

Every prime number greater than 3 is either 1 less or 1 more than a multiple of 6.

$$
\begin{aligned}
5 &= 6 - 1\\
7 &= 6 + 1\\
11 &= 12 - 1\\
13 &= 12 + 1\\
17 &= 18 - 1\\
19 &= 18 + 1\\
23 &= 24 - 1\\
29 &= 30 - 1\\
31 &= 30 + 1\\
37 &= 36 + 1\\
41 &= 42 - 1\\
43 &= 42 + 1\\
47 &= 48 - 1\\
53 &= 54 - 1\\
&\vdots
\end{aligned}
$$

NUMBER THEORY TRIVIA

When looking at a large table of prime numbers you may notice the pattern

31
331
3331
33331
333331
3333331
33333331

Patterns such as this look like they might produce prime numbers indefinitely, but eventually, as always seems to be the case, the pattern ceases to produce primes. In this case, the next number, 333,333,331 is exactly divisible by 17.

15. Write a program that will generate primes between 2000 and 3000.

16. To test whether a number n is prime or not you only have to show that it is not divisible by any number less than or equal to the square root of n. Why?

17. There are only seven prime number years in the period 1950-2000. What are the prime number years, if any, from 1980 to 1989?

18. Is the number 961 a prime number? Find out by using the Prime Check program.

19. What is the largest prime you need to consider to be sure that you have excluded, in the Sieve of Eratosthenes, all primes less than or equal to:
    a. 300 b. 2000 c. 5000

20. Modify the Sieve program so that it will produce prime numbers less than 2000.

21. Modify the Wilson's Theorem program so that it will compute a table of values for the integers 2 through 20.

22. Modify the Rare Prime Decades program so that it will compute a list of rare primes up to 10,000.

23. Prime numbers of the form $2^p - 1$, where p is a prime number, are called ______________ .

24. Explain the difference between Mersenne numbers and Mersenne primes.

25. Who discovered the 30th Mersenne prime?

26. Who discovered the 31st Mersenne prime?

27. What computer was used in the generation of the 30th Mersenne prime?

28. How many twin primes are found in the list of primes from 1 to 1000?

29. Triplet primes are three primes of the form n, n + 2, n + 4. Give an example of triplet primes.

30. What are pairs of primes called?

31. Two consecutive odd prime numbers, such as 5 and 7, or 17 and 19, are called twin primes. The existence or nonexistence of an infinite number of such number pairs is still one of the unsolved problems of number theory. The number of prime triplets, such as 3, 5, 7, however, has been proven to be finite. Prime triplets are three primes of the form n, n + 2, n + 4. Write a program that will produce several prime triplets.

32. Modify the Twin Primes program so that it will generate all pairs of twin primes less than 2000.

NUMBER THEORY TRIVIA

The reverse of the prime 23 is 32 which equal 2^5. Other primes of this form are:

PRIME	REVERSE OF PRIME	POWER
521	125	5^3
821	128	2^7
4201	1024	2^{10}
270131	131072	2^{17}
806041	140608	52^3

Can you write a program to generate additional primes of this form?

NUMBER THEORY TRIVIA

Truncatable primes are primes (with at least two digits and without the digit zero) that remain prime no matter how many of the left-most digits are deleted. For example, the prime number 4632647 is truncatable because each of its truncations: 632647, 32647, 2647, 647, 47, and 7 are prime. Other truncatable primes are:

	36484	95721	35366	76883
3	15396	33424	56637	86197
99	18918	99765	33196	93967
966	86312	64621	65676	29137

Can you write a program to produce other truncatable primes?

33. Does the formula $X(n) = 3^n - 1$ give primes for n = 1, 2, 3, 4?
34. Is there a formula that will give all the prime numbers?
35. For what values of n is $11 \times 14^n + 1$ a prime?
36. The polynomial $x^2 + x + 17$ may be used to generate sixteen prime numbers when x = 0, 1, 2, . . ., 15. Write a program that will produce a printout of these sixteen prime numbers.
37. Useless formulas for generating one or a couple of primes are easily constructed. Examples of such formulas are $2 + 1^n$ and $2 + n/n$. Find some other useless formulas that generate primes.
38. Write a program that will generate primes using the formula
$$9n^2 - 489n + 6683$$
39. What is an emirp? Give an example.

NUMBER THEORY TRIVIA

Look at the 6-digit prime

193939

It is not only reversible, but also cyclic. By successively moving the initial digit to the end, one obtains the series 193939, 939391, 393919, 939193, 391939 and 919393. One more such move restores the original order. Each one of the six numbers is a reversible prime.

3

INTERESTING NUMBERS

PREVIEW

Since the time of Pythagoras, the study of numbers has been pursued by noted mathematicians and interested amateurs the world over. Numbers are fascinating entities. Consider the number "17": does it exist in nature? There are collections of things that contain seventeen objects; you may have seventeen quarters in your pocket, or walk up seventeen flights of stairs. But those quarters aren't "17" anymore than they are George Washington. The number "17" goes far beyond any specific embodiment of it. Seventeen has properties independent of whether one is considering quarters, or stairs, or stars. In counting upwards, 17 follows 16 and precedes 18. Seventeen is a "prime number," that is, it has no factors except itself and 1 among the positive integers. It is possible to construct a seventeen-sided regular polygon using only a compass and a straightedge, just as it's possible to make an equilateral triangle or pentagon. Seventeen is very special, in lots of ways; so are other numbers. A variety of interesting numbers and number patterns are discussed in this chapter.

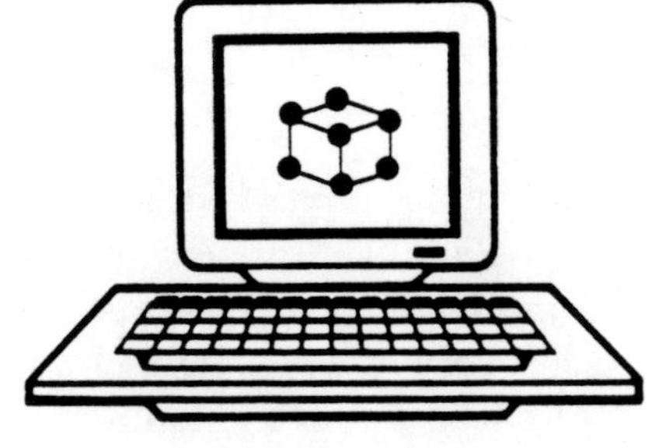

After you finish this chapter, you should be able to:

1. Identify perfect numbers, multiply perfect numbers, multiperfect numbers, amicable numbers, Armstrong numbers, tetranacci numbers, Lucas numbers, tribonacci numbers, lucky numbers, tetrahedral numbers, abundant numbers, deficient numbers, triangular numbers, rectangular numbers, square numbers, star numbers, and powerful numbers.

Leonhard Euler (1707-1783), one of the greatest mathematicians of all time, knew of only four perfect numbers: 6, 28, 496, and 8128. Euler was born in Switzerland. He wrote in all mathematical fields, creating new results as well as organizing several fields, above all calculus and analysis. He possessed a phenomenal memory, had almost total recall, and could mentally calculate long and complicated problems.

2. See how a microcomputer can be used to generate the above numbers.
3. Identify several formulas and techniques that can be used to produce the above numbers.
4. Write programs to produce some of the interesting numbers found in the number theory arena.

3.1 PERFECT NUMBERS

A number is said to be perfect if it is the sum of the divisors other than itself. Thus 6 and 28 are perfect numbers since 6 = 3 + 2 + 1, 28 = 14 + 7 + 4 + 2 + 1. The first four perfect numbers are 6, 28, 496, and 8128. The ancient Greek and early Judaic philosophers knew about these four perfect numbers and endowed them with mystical properties. They tried, but were unable to find an odd perfect number. Euclid recognized - and only the Greek gods know how - that the first four perfect numbers are generated by the formula

$$2^{n-1}(2^n - 1)$$

for n equal to 2, 3, 5 and 7. The computations are as follows:

$$2^1(2^2 - 1) = 2(3) = 6$$
$$2^2(2^3 - 1) = 4(7) = 28$$
$$2^4(2^5 - 1) = 16(31) = 496$$
$$2^6(2^7 - 1) = 64(127) = 8128$$

Euclid saw that in all four computations, $2^n - 1$ was a prime (3, 7, 31 and 127). This observation inspired him to prove a powerful theorem: the formula

$$2^{n-1}(2^n - 1)$$

generates an even perfect number whenever $2^n - 1$ is a prime. The ancients also observed that perfect numbers ended in a 6 or an 8.

Fourteen centuries after Euclid, the fifth perfect number was discovered, and it contained eight digits: 33,550,336.

Euclid's proof that

$$2^{n-1}(2^n - 1)$$

will yield a perfect number whenever $2^n - 1$ is a prime says nothing about which integral values of n will make $2^n - 1$ a prime. The truth is that n must be a prime for $2^n - 1$ to be a prime, but n's being prime does not in itself make $2^n - 1$ a prime. Indeed, for most prime values of n, $2^n - 1$ is not a prime.

Numbers generated by the expression $2^n - 1$ are known as Mersenne numbers, after a seventeenth-century Parisian monk, Marin Mersenne, who took time from his monastic duties for number theory. On account of Euclid's formula, every time a new prime Mersenne number is discovered, a new perfect number is

NUMBER THEORY TRIVIA

139 and 149 are the first consecutive primes differing by 10.

NUMBER THEORY TRIVIA

There is a one-to-one correspondence between **even perfect numbers** and **Mersenne primes.** Euclid proved that $2^{n-1}(2^n - 1)$ is perfect whenever $2^n - 1$ is prime. Since, as proved by Euler, all even perfect numbers have this form, the question of the infinitude of the set of even perfect numbers is equivalent to the same question for the set of Mersenne primes, i.e., primes of the form $2^n - 1$.

automatically known. In 1644, Mersenne himself stated that the Mersenne numbers $2^{13} - 1$, $2^{17} - 1$ and $2^{19} - 1$ are primes (8191; 131,071; and 524,287).

The monk also claimed that $2^{67} - 1$ would prove to be a prime. This bold claim went unchallenged for more than a quarter of a millennium. In 1903, at a meeting of the American Mathematical Society, Frank Cole, a Columbia University professor, rose to deliver a paper modestly titled "On the Factorization of Large Numbers." Cole, who was a man of very few words, walked to the board and, saying nothing, proceeded to chalk up the arithmetic for raising 2 to the sixty-seventh power. Then he carefully subtracted 1. Without a word he moved over to clear space on the board and multiplied out . . . 193,707,721 × 761,838,257,287. The two numbers agreed. Mersenne's intrepid claim about $2^{67} - 1$ vanished into the limbo of mathematical mythology. For the first time and only time on record, an audience of the American Mathematical Society vigorously applauded the author of a paper delivered before it. Cole took his seat without having uttered a word. Nobody asked him a question.

Some 2000 years after Euclid proved that his formula always yields even perfect numbers, the eighteenth-century Swiss mathematician Leonhard Euler proved that the formula will yield all the even perfect numbers.

In recent years, computers have been used to generate very large perfect numbers: the thirtieth perfect number contains 130,000 digits:

$$2^{216090}(2^{216091} - 1)$$

For more than 2500 years mathematicians have known that there are infinitely many primes. But in that much time, they have not been able to determine whether perfect numbers are also inexhaustible.

Are there any odd perfect numbers? In 1985, Michael Friedman, using an IBM Personal Computer, demonstrated that there are no odd perfect numbers below 10^{79} that have eight prime divisors (which is the minimum number of prime divisors an odd perfect number could have).

The quest for perfect numbers is still continuing. Where will it lead? What will it accomplish? The next perfect number will take a bookshelf of books to write it out in full, the one after that a small library of books, the one after that all the books in the Library of Congress, the one after that all the books in the libraries in the United States, and the one after that all the books in the world. Why scale Mt. Everest; why reach the moon? Are there infinitely many perfect numbers? It might take another 2500 years to know. Perfect number is the right term for these numbers; only the perfect could be so rare.

LARGEST
PERFECT NUMBER

- $2^{216090}(2^{216091} - 1)$
- 130,100 DIGITS LONG

The following program computes the first four perfect numbers. The first two numbers are computed immediately. It will take the microcomputer about a half hour to compute the third number and several hours of computing before the fourth number is found.

```
program PerfectNumbers;
var
  D,N,S       :Integer;
begin
  for N := 2 to 8200 do
    begin
      S := 0;
        for D := 1 to N div 2 do
        if N div D = N/D then S := S + D;
      if S = N then writeln(N, ' IS A PERFECT NUMBER');
    end;
end.
```

COMPILE/RUN

```
6 IS A PERFECT NUMBER
28 IS A PERFECT NUMBER
496 IS A PERFECT NUMBER
8128 IS A PERFECT NUMBER
```

The following program can be used to determine if any given number N is perfect.

```
program PerfectNumberCheck;
var
  N          :Integer;(* Number *)
  S,D        :Integer;(* Calculating variables *)
begin
  write('ENTER THE NUMBER ');
  readln(N);
  D := 1;
  S := 0;
  repeat
    if N/D < = N div D then S := S + D;
    D := D + 1;
  until D > N/2;
  if S = N
    then writeln(N,' IS A PERFECT NUMBER ')
    else writeln(N,' IS NOT A PERFECT NUMBER');
end.
```

COMPILE/RUN

```
ENTER THE NUMBER 286
286 IS NOT A PERFECT NUMBER
```

I didn't know any number was perfect.

NUMBER THEORY TRIVIA

No **odd perfect number** has ever been found and the question of the existence of an odd perfect number has become one of the most notorious problems of number theory. If an odd perfect number exists then it must be greater than 10^{50} and have at least 8 distinct prime factors.

Perfect Number	Number of Digits
1. $2^1(2^2 - 1) = 6$	1
2. $2^2(2^3 - 1) = 28$	2
3. $2^4(2^5 - 1) = 496$	3
4. $2^6(2^7 - 1) = 8{,}128$	4
5. $2^{12}(2^{13} - 1) = 33{,}550{,}336$	8
6. $2^{16}(2^{17} - 1) = 8{,}589{,}869{,}056$	10
7. $2^{18}(2^{19} - 1) = 137{,}438{,}691{,}328$	12
8. $2^{30}(2^{31} - 1)$	19
9. $2^{60}(2^{61} - 1)$	37
10. $2^{88}(2^{89} - 1)$	54
11. $2^{106}(2^{107} - 1)$	65
12. $2^{126}(2^{127} - 1)$	77
13. $2^{520}(2^{521} - 1)$	314
14. $2^{606}(2^{607} - 1)$	366
15. $2^{1278}(2^{1279} - 1)$	770
16. $2^{2202}(2^{2203} - 1)$	1,327
17. $2^{2280}(2^{2281} - 1)$	1,373
18. $2^{3216}(2^{3217} - 1)$	1,937
19. $2^{4252}(2^{4253} - 1)$	2,561
20. $2^{4422}(2^{4423} - 1)$	2,663
21. $2^{9688}(2^{9689} - 1)$	5,834
22. $2^{9940}(2^{9941} - 1)$	5,985
23. $2^{11212}(2^{11213} - 1)$	6,751
24. $2^{19937}(2^{19938} - 1)$	12,003
25. $2^{21700}(2^{21701} - 1)$	13,066
26. $2^{23208}(2^{23209} - 1)$	13,973
27. $2^{44496}(2^{44497} - 1)$	26,790
28. $2^{86242}(2^{86243} - 1)$	51,924
29. $2^{110502}(2^{110503} - 1)$	—
30. $2^{132048}(2^{132049} - 1)$	79,502
31. $2^{216090}(2^{216091} - 1)$	130,100

The first thirty-one perfect numbers.

3.2 MULTIPERFECT NUMBERS

When the sum of the divisions of an integer n is a multiple of n, we call n a multiperfect number. For example, 28 is a multiperfect number since the sum of its divisors is twice the number: that is,

$$1 + 2 + 4 + 7 + 14 + 28 = 56 = 2 \times 28$$

NUMBER THEORY TRIVIA

n	Number of primes less than 10^n
1	4
2	25
3	168
4	1229
5	9592
6	78,498
7	644,579
8	5,761,455
9	50,847,534
10	455,052,511

Multiperfect numbers are also called multiply perfect numbers. The following program finds all the multiperfect numbers less than 1000.

```
program MultiperfectNumbers;
var
  N      :Integer;
  D1     :Integer;(* Divisor of N *)
  S      :Integer;(* Sum of the divisors *)
begin
  writeln('INTEGER        SUM OF DIVISORS        MULTIPLE');
  writeln('------------------------------------------------------');
```

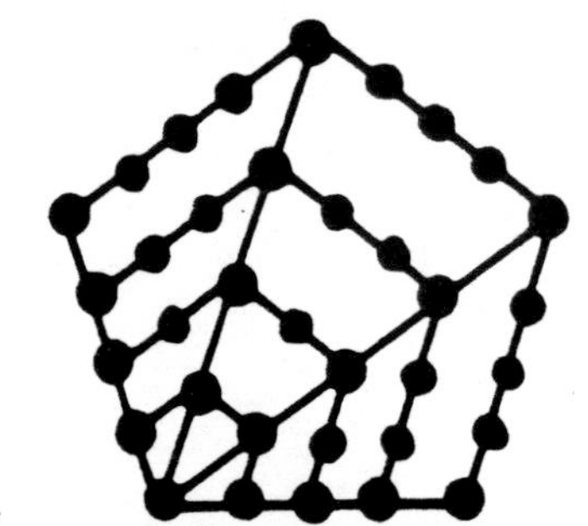

Pentagonal numbers can be computed using the formula $n/2(3n - 1)$. The first few numbers are

1 5 12 22 35 51 70 92 117

Can you write a program to generate the first fifty pentagonal numbers?

```
  for N := 1 to 1000 do
    begin
      S := 0;
      (* Find sum of the divisors of N *)
      for D1 := 1 to TRUNC(SQRT(N)) do
        if N div D1 = N / D1 then S := S + D1 + N div D1;
      (**** Is the sum of the divisors a multiple of N? ****)
      if (S / N = S div N) and (N <> 1) then
        writeln(N:4,'          ',S:4,'          ',S div N:2);
    end;
end.
```

COMPILE/RUN

INTEGER	SUM OF DIVISORS	MULTIPLE
1	1	1
6	12	2
28	56	2
120	360	3
496	992	2
672	2016	3

There is no known rule or formula that will yield multiperfect numbers. Much work remains to be done with multiperfect numbers. Among the many questions still requiring answers are: Are there new multiperfect numbers that have not been discovered? Are there any odd multiperfect numbers? Are there any multiperfect numbers beyond the octoperfect class (the sum of all the divisors is equal to 8 times the integer itself)?

3.3 AMICABLE NUMBERS

Once there was a king who thought of himself as quite a mathematician. He told a prisoner, "Give me a problem to solve. You may go free until I solve it. But as soon as I have the answer, off comes your head!" Now the prisoner was rather clever himself. Here is the problem he gave the king. 220 and 284 are called amicable numbers (also called friendly numbers). The sum of the proper divisors of 220 equals 284,

$$1 + 2 + 4 + 5 + 10 + 11 + 20 + 22 + 44 + 55 + 110 = 284$$

and the sum of the proper divisors of 284 equals 220.

$$1 + 2 + 4 + 71 + 142 = 220$$

Find the next pair of amicable numbers! The story goes that the prisoner went free and finally died of old age. The king never solved the problem.

NUMBER THEORY TRIVIA

With the aid of computers, all the **amicable number pairs** up to 100 million have now been determined.

NUMBER THEORY TRIVIA

A **reverse square** is shown below:

$12^2 = 144$

and

$21^2 = 441$

Can you write a program to generate other reverse squares?

Some of the amicable pairs of numbers are:

1184	1210
2620	2924
5020	5564
6232	6368
10744	10856
12285	14595
17296	18416
63020	76084
66928	66992
67095	71145
69615	87633
79750	87730
9,363,584	9,437,056
111,448,537,712	118,853,793,424

Several methods are available for finding amicable pairs. One method is to let

$$A = (3)(2^x) - 1$$
$$B = (3)(2^{x-1}) - 1$$
$$C = (9)(2^{2x-1}) - 1$$

If x is greater than 1, and A, B, and C are all primes, then 2^xAB and 2^xC constitute an amicable pair of numbers. For example, if x = 4, then A = 47, B = 23, and C = 1151, which are all primes. Then

$$(2)^4(47)(23) = 17296$$

and

$$(2^4)(1151) = 18416$$

Too bad the king didn't have a computer. He could have written the following program to produce the next pair of amicable numbers.

```
program AmicableNumbers;
var
  A,B,C,D,S,T,F        :Integer;
begin
  for A := 1 to 7000 do
    begin
      S := 0;
      for D := 1 to A div 2 do
        if A / D = A div D then S := S + D;
      if S > A then
        begin
          B := S;
          T := 0;
          for F := 1 to B div 2 do
```

```
                    if B / F = B div F then T := T + F;
                if T + A then writeln(A, ' AND ', B, ' ARE AMICABLE NUMBERS');
            end;
        end;
end.
```

COMPILE/RUN

220 AND 284 ARE AMICABLE NUMBERS
1184 AND 1210 ARE AMICABLE NUMBERS

There are a total of five pairs of amicable numbers less than 10,000. A scheme for finding them is as follows:

1. Start with a number, N
2. Factor it, and find the sum, S, of its factors, counting 1 as a factor.
3. If S is more than N, find S1, the sum of the factors of S.
4. If S1 = N, print the amicable pair: N, S.
5. Only even values of N are used, since all amicable numbers are even.

Let's use this scheme on the pair: 220, 284.

1. N = 220
2. Sum of factors of 220 is 284. Thus, S = 284.
3. Since S(284) is more than N(220), find S1 (sum of factors of 284). S1 = 220.
4. S1(220) = N(220). Therefore, 220 and 284 are amicable numbers.

The reader should write a program using this procedure.

The search for amicable pairs is eminently suited to computers. For each number N let the computer determine all divisors ($\neq$ N) and their sum M. Then perform the same operation on M. If you return to the original number, N, by this procedure, an amicable pair (N, M) has been discovered. Several years ago, this procedure was used on a computer with N's of up to one million resulting in the collection of 42 pairs of amicable numbers.

Actually we know very little about the properties of the amicable numbers, but on the basis of the numbers shown one can make some conjectures. For instance, it appears that the quotient of the two numbers must get closer and closer to 1 as they increase. From the list of numbers given you can see that both numbers may be even or both odd, but no case has been found in which one was odd and the other even.

There are, besides amicable pairs, numbers that are amicable triplets. That is, an amicable triplet consists of 3 numbers so related

NUMBER THEORY TRIVIA

The largest perfect square whose digits are in strictly increasing order is

$$134689 = 367^2$$

Can you write a program to produce other perfect squares of this type?

NUMBER THEORY TRIVIA

Numbers like 153 and 407 are called **narcissistic numbers**.

$$153 = 1^3 + 5^3 + 3^3$$
$$407 = 4^3 + 0^3 + 7^3$$

An impressive narcissistic number is

$$165{,}033 = 16^3 + 50^3 + 33^3$$

that the sum of the proper divisors of any one of them is equal to the sum of the other 2 numbers.

One such amicable triplet is: 103,340,640; 123,228,768; and 124,015,008. Another amicable triplet is: 1,945,330,728,960; 2,324,196,638,720; and 2,615,631,953,920. These are not easy to find. The numbers in this set have 959, 959, and 479 divisors, respectively. Perhaps the reader would like to use a microcomputer to find a few amicable triplets!

3.4 ARMSTRONG NUMBERS

One hundred fifty-three is an interesting number because

$$153 = 1^3 + 5^3 + 3^3$$

Numbers such as this are called Armstrong numbers. Any N digit number is an Armstrong number if the sum of the Nth power of the digits is equal to the original number.

The following program finds three other three-digit Armstrong numbers.

```
program ArmstrongNumbers;
var
  A,B,C,N        :Integer;
begin
  for N := 100 to 999 do
    begin
      A := N div 100;
      B := N div 10 - 10 * A;
      C := N - 100 * A - 10 * B;
      if N = A*A*A + B*B*B + C*C*C then
        begin
          writeln('ARMSTRONG NUMBER ', N);
          writeln('EQUALS ', A*A*A, ' + ', B*B*B, ' + ', C*C*C;
          writeln;
        end;
    end;
end.
```

COMPILE/RUN

```
ARMSTRONG NUMBER  153
EQUALS 1 + 125 + 27

ARMSTRONG NUMBER 370
EQUALS 27 + 343 + 0

ARMSTRONG NUMBER  371
EQUALS 27 + 343 + 1

ARMSTRONG NUMBER  407
EQUALS 64 + 0 + 343
```

NUMBER THEORY TRIVIA

A **Copperbach number** is a number that can be represented as a sum of two prime numbers in two different ways.

$$18 = 5 + 13$$
$$= 7 + 11$$

$$20 = 17 + 3$$
$$= 13 + 7$$

Can you write a program to generate several Copperbach numbers?

1
1
2
4
8
15
29
56
108
208
401
773
1490
2872
5536
10671
20569
39648
76424
147312
283953
547337
1055026
2033628
3919944
7555935

The first twenty-six tetranacci numbers.

3.5 TETRANACCI NUMBERS

The tetranacci numbers are an infinite sequence of numbers - starting with 1, 1, 2, 4, 8, 15, 29, 56, 108, 208 - each of which (after the fourth number) is the sum of the four preceding it. For example, 8 is the sum of 1, 1, 2, and 4; 15 is the sum of 1, 2, 4 and 8.

The following program generates the first 17 tetranacci numbers.

```
program TetranacciNumbers;
var
  T      :array[1..17] of Integer;
  N      :Integer;(* Counter and array index *)
  X      :Integer;(* Counter for printout *)
begin
  (**** Establish first four Tetranacci numbers ****)
  T[1] := 1;
  T[2] := 1;
  T[3] := 2;
  T[4] := 4;
  (**** Compute last 13 Tetranacci numbers ****)
  for N := 1 to 13 do
    T[N+4] := T[N+3] + T[N+2] + T[N+1] + T[N];
  (**** Print contents of array T ****)
  writeln('TETRANACCI NUMBERS');
  writeln;
  for X := 1 to 17 do
    writeln(T[X]);
end.
```

3.6 LUCAS NUMBERS

A sequence of numbers is named after the French mathematician Francois Edouard Anatole Lucas (1842-1891). The terms in the Lucas sequence are

1, 3, 4, 7, 11, 18, 29, 47, 76, . . .

A number in the sequence is the sum of the two preceding it. The number of Lucas numbers is unlimited.

The following program will compute the first 21 Lucas numbers.

```
program LucasNumbers;
var
  F      :array[1..2] of Integer;
  I      :Integer;
begin
  (**** Initialize first two Lucas numbers ****)
  F[1] := 1;
  F[2] := 3;
```

NUMBER THEORY TRIVIA

396,733 and 396,833 are the first consecutive primes differing by 100.

```
  (**** Compute Lucas numbers ****)
  for := 3 to 21 do
    F[I] := F[I-2] + F[I-1];
  (**** Print Lucas numbers ****)
  writeln('LUCAS NUMBERS');
  ('---------------------------');
  for I := 1 to 21 do
    writeln(F[I]);
end.
```

```
COMPILE/RUN

LUCAS NUMBERS
--------------------
1
3
4
7
11
18
29
47
76
123
199
322
521
843
1364
2207
3571
5778
9349
15127
24476
```

NUMBER THEORY TRIVIA

A **trimorphic number** is one that terminates its associated triangular number (a triangular number is of the form $T(n) = n(n+1)/2$. There are six trimorphic numbers less than 100,000. Five of them are

n	T(n)
1	1
5	15
25	325
625	195625
9376	43959376

Can you write a program to find the sixth number?

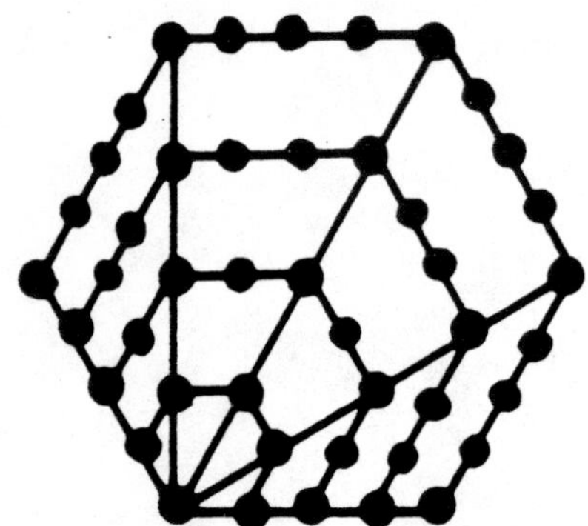

Hexagonal numbers can be computed using the formula $n(2n - 1)$. The first few numbers are

1 6 15 28 45 66

Can you write a program to generate the first fifty hexagonal numbers.?

3.7 TRIBONACCI NUMBERS

The tribonacci number sequence is the sequence 1, 1, 2, 4, 7, . . . in which each new term, excepting the first three terms, is the sum of the preceding three terms. The number of tribonacci numbers is unlimited.

The following program produces the first eighteen tribonacci numbers.

```
program TribonacciNumbers;
var
  T       :array[1..20] of Integer;
  I       :Integer;
begin
  (**** Initialize first three Tribonacci numbers ****)
  T[1] := 1;
  T[2] := 1;
  T[3] := 2;
  (**** Compute Tribonacci numbers ****)
  for I := 4 to 18 do
    T[I] := T[I-3] + T[I-2] + T[I-1];
  (**** Print Tribonacci numbers ****)
  writeln('TRIBONACCI NUMBERS');
  writeln('------------------------');
  for I := 1 to 18 do
    writeln(T[I]);
end.
```

COMPILE/RUN

```
TRIBONACCI NUMBERS
------------------------
1
1
2
4
7
13
24
44
81
149
274
504
927
1705
3136
5768
10609
19513
```

NUMBER THEORY TRIVIA

Pentagonal numbers follow the pattern

1 = 1
1+4 = 5
1+4+7 = 12
1+4+7+10 = 22
1+4+7+10+13 = 35
1+4+7+10+13+16 = 51
1+4+7+10+13+16+19 = 70

Of course I flunked spelling! Why didn't she give me plain, ordinary words like tribonacci, tetranacci, multiperfect, tetrahedral, or pentagonal?

NUMBER THEORY TRIVIA

A pentagonal number has the form x(3x - 1)/2. A **palindromic pentagonal number** is one that reads the same forwards as it does backwards. The following table lists several palindromic pentagonal numbers.

x	x(3x - 1)/2
1	1
2	5
4	22
26	1001
44	2882
101	15251
693	720027
2173	7081807
2229	7451547
4228	26811862
6010	54177145
26466	1050660501
26906	1085885801
31926	1528888251
44059	2911771192
1258723	2376574756732
1965117	5792526252975
1979130	5875432345785
2684561	10810300301801

3.8 LUCKY NUMBERS

Lucky numbers were discovered by Stanislav Ulam and a group of scientists at Los Alamos Scientific Laboratories around 1956. These numbers have many properties that are very similar to those of the prime numbers. Lucky numbers are found by a sieve process similar to the sieve method the Greek Eratosthenes used to find prime numbers.

As with the sieve of Eratosthenes, we begin by writing down all the natural numbers, in order. If we leave 1 and strike out every second number, we eliminate all the even numbers. In Eratosthenes' sieve we next struck out every multiple of 3 because 3 was the next surviving number. The rule here is different: strike out every third number among those remaining. That means that 5 goes, and 11, 17, 23, etc. The next surviving number is 7, so we let that stand and cross out every seventh remaining one (19, 39, etc.). Then cross out every 9th, then every 13th, and so on. The first 56 lucky numbers are

1	31	69	111	159	205	259
3	33	73	115	163	211	261
7	37	75	127	169	219	267
9	43	79	129	171	223	273
13	49	87	133	189	231	283
15	51	93	135	193	235	285
21	63	99	141	195	237	289
25	67	105	151	201	241	297

The following program verifies by the sieve method that the above list of lucky numbers is correct.

```
program LuckyNumbers;
var
  A          :array[1..300] of Integer;
  I,C,X,B    :Integer;
begin
  (**** Fill array A with 1's ****)
  for I := 1 to 297 do
    A[I] := 1;
  (**** Find Lucky numbers ****)
  C := 0;
  repeat
    repeat
      C := C + 1;
    until A[C] <> 0;
    X := C;
    if C = 1 then X := 2;
    B := 0;
    for I := 1 to 297 do
```

```
            if A[I] <> 0 then
              begin
                B := B + 1;
                if B >= X then
                  begin
                    A[I] := 0;
                    B := 0;
                  end;
              end;
      until C >= 296;
      (**** Print Lucky numbers ****)
      for I := 1 to 297 do
        if A[I] <> 0 then writeln(I,'      ');
    end.
```

I call them unlucky numbers because they are the numbers of the games that I lost in the arena last Sunday.

NUMBER THEORY TRIVIA

A tetrahedral number has the form x(x + 1)(x + 2)/6. A **palindromic tetrahedral number** is one that reads the same forwards as it does backwards. The following table lists five palindromic tetrahedral numbers.

x	x(x+1)(x+2)/6
1	1
2	4
17	969
21	1771
336	6378736

3.9 TETRAHEDRAL NUMBERS

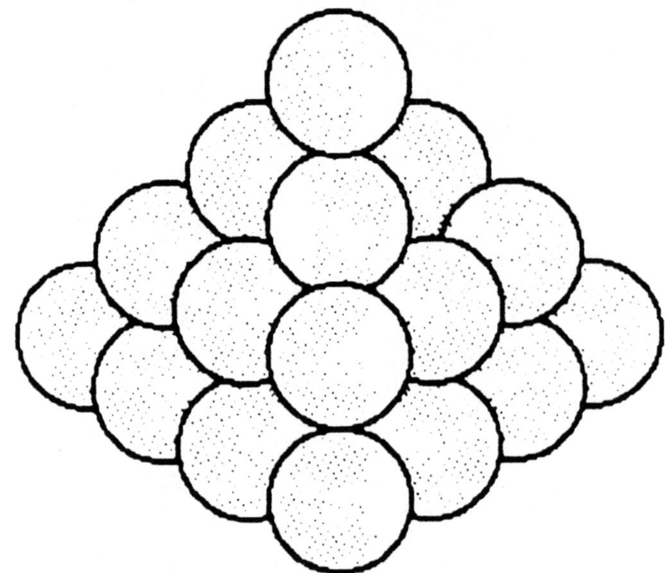

The grounds in front of the fort at St. Augustine, Florida is decorated with cannons and cannon balls piled in the form of a pyramid. A pyramidal piling of cannon balls, or any other objects, give rise to different types of pyramidal numbers. Tetrahedral or triangular pyramidal numbers are the simplest kind of pyramidal numbers. The first four tetrahedral numbers are 1, 4, 10, and 20. There are an unlimited number of tetrahedral numbers.

The following program computes the first 30 tetrahedral numbers.

```
program TetrahedralNumbers;
var
  T,X       :Integer;
begin
  writeln('TETRAHEDRAL  NUMBERS');
  writeln('----------------------------');
  for X := 1 to 30 do
    begin
      T := (X * (X + 1) * (X + 2)) div 6;
      writeln(T);
    end;
end.
```

COMPILE/RUN

```
TETRAHEDRAL  NUMBERS
----------------------------
1
4
10
20
35
56
84
```

NUMBER THEORY TRIVIA

A **Silverbach number** is a number that can be represented as the sum of two primes in three different ways.

$$22 = 19 + 3$$
$$= 17 + 5$$
$$= 11 + 11$$

Can you write a program to generate several Silverbach numbers?

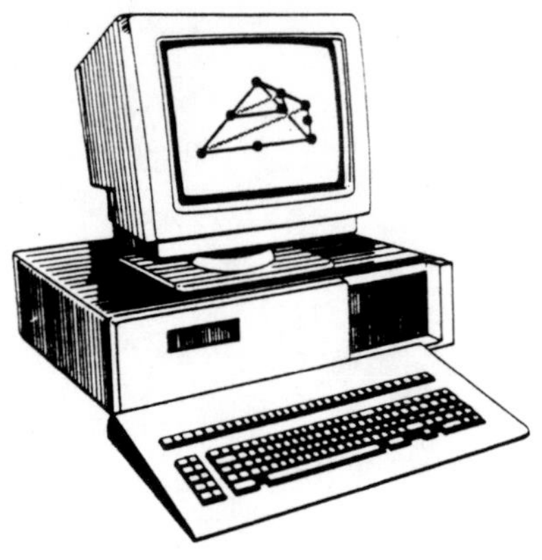

120
165
220
286
364
455
560
680
816
969
1140
1330
1540
1771
2024
2300
2600
2925
3276
3654
4060
4495
4960

3.10 ABUNDANT AND DEFICIENT NUMBERS

The number, the sum of whose divisors is less than the number itself, is called deficient, and a number exceeded by this sum is called abundant. The number is perfect when the sum of the divisors of that number, excluding the number itself, equals the number in question.

For example,

$6 = 1 + 2 + 3$ and is perfect
$12 < 1 + 2 + 3 + 4 + 6$ and is abundant.
$10 > 1 + 2 + 5$ and is deficient.

The following program factors a given number into its divisors and determines whether the number is abundant, deficient, or perfect.

```
program AbundantAndDeficientNumbers;
var
  N      :Integer;
  X      :Integer;(* Divisor of number *)
  S      :Integer;(* Sum of the divisors *)
  Z      :Integer;(* Continue/Stop indicator *)
begin
  (**** This program will take a number and ****)
  (**** compute the sum of its divisors ****)
  write('ENTER THE NUMBER: ');
  readln(N);
```

NUMBER THEORY TRIVIA

Hexagonal numbers follow the pattern

1	= 1
1+5	= 6
1+5+9	= 15
1+5+9+13	= 28
1+5+9+13+17	= 45
1+5+9+13+17+21	= 66

NUMBER THEORY TRIVIA

Triangular numbers are part of a large classification of numbers known as figurate numbers, that is, numbers related in a special way to certain geometric figures — triangles, squares, pentagons, and so on. Figurate numbers, and especially the triangular ones, were viewed with mystical reverence by the ancient Greeks, most notably the followers of Pythagoras. For example, the triangular number ten was considered to be "perfection," because, among other things, it was the sum of 1 (the point), 2 (the line), 3 (the plane), and 4 (the solid). Hence, their union, that is, all space, was the number of the decade itself. Even when the magic or religion of such numbers faded with the passing of time, figurate numbers still held a fascination for mathematicians interested in number theory.

```
  S := 0;
  write('THE DIVISORS OF ',N,' ARE ');
  for X := 1 to N - 1 do
    if N/X = N div X then
      begin
        S := S + X;
        write(X:6);
      end;
  writeln;
  if S > N then writeln(N, ' IS ABUNDANT');
  if S < N then writeln(N, ' IS DEFICIENT');
  if S = N then writeln(N, ' IS PERFECT');
end.
```

COMPILE/RUN

```
ENTER THE NUMBER: 15
THE DIVISORS OF 15 ARE 1 3 5
15 IS DEFICIENT

ENTER THE NUMBER: 28
THE DIVISORS OF 28 ARE 1 2 4 7 14
28 IS PERFECT

ENTER THE NUMBER: 12
THE DIVISORS OF 12 ARE 1 2 3 4 6
12 IS ABUNDANT
```

3.11 TRIANGULAR NUMBERS

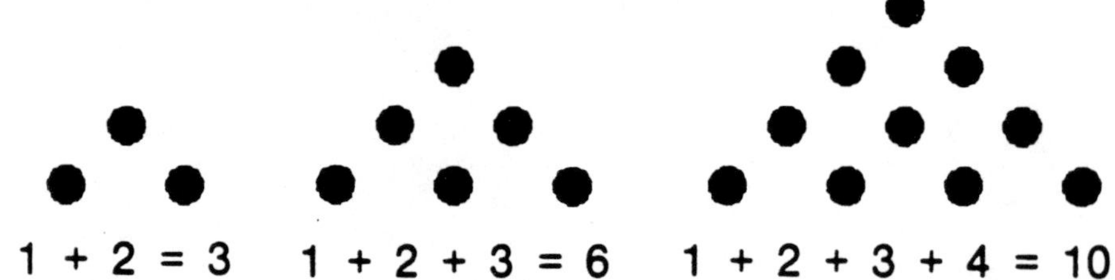

1   1 + 2 = 3   1 + 2 + 3 = 6   1 + 2 + 3 + 4 = 10

Triangular numbers are consecutive numbers that form triangles. To find the sum of any group of consecutive numbers, the formula

$$\text{Sum} = N(N + 1) / 2$$

can be used. N is the value of the last consecutive number and the first number is 1. For example, the sum of the first six numbers is

$$6(6 + 1) / 2 = 42 / 2 = 21$$

NUMBER THEORY TRIVIA

1,000,000,000,061
and
1,000,000,000,063
are large twin primes.

1
3
6
10
15
21
28
36
45
55
66
78
91
105
120
136
153
171
190
210
231
253
276
300
325
351
378
406
435
465
496
528
561
595
630
666
703
741
780
820
861
903
946
990
1035
1081
1128
1176
1225
1275

The first fifty triangular numbers.

The following program will calculate triangular numbers.

```
program TriangularNumbers;
var
  L,N,T      :Integer;
begin
  write('ENTER NUMBER OF TRIANGULAR NUMBERS: ');
  readln(L);
  writeln;
  writeln('TERM               TRIANGULAR NUMBER');
  writeln('-----------------------------------------');
  for N := 1 to L do
    begin
      T := (N * (N + 1)) div 2;
      writeln(N:3, '                 ', T:4);
    end;
end.
```

Let's consider another idea about triangular numbers. It is possible to find a pattern that will let us identify the 19th or the Nth triangular number without producing all the preceding numbers in the series up to those points. To find the 7th number, we simply multiply 7 × 8 and divide the product by two; to find the 19th number, we multiply 19 × 20 and divide the product by 2. Is 97 a triangular number? Is 210?

The following program will determine if a specific number is a triangular number.

```
program TriangularNumberCheck;
var
  N      :Integer;(* Number *)
  R      :Real;(* Calculating variable *)
  T      :Real;(* Triangular number if T = integer *)
begin
  write('ENTER A NUMBER ');
  readln(N);
  R := SQRT(1 + 8 * N);
  T := (-1 + R) / 2;
  (**** N is triangular if T is an integer ****)
  if T = INT(T)
    then writeln(N,' IS A TRIANGULAR NUMBER')
    else writeln(N,' IS NOT A TRIANGULAR NUMBER');
end.
```

```
COMPILE/RUN

ENTER A NUMBER 55
55 IS A TRIANGULAR NUMBER
```

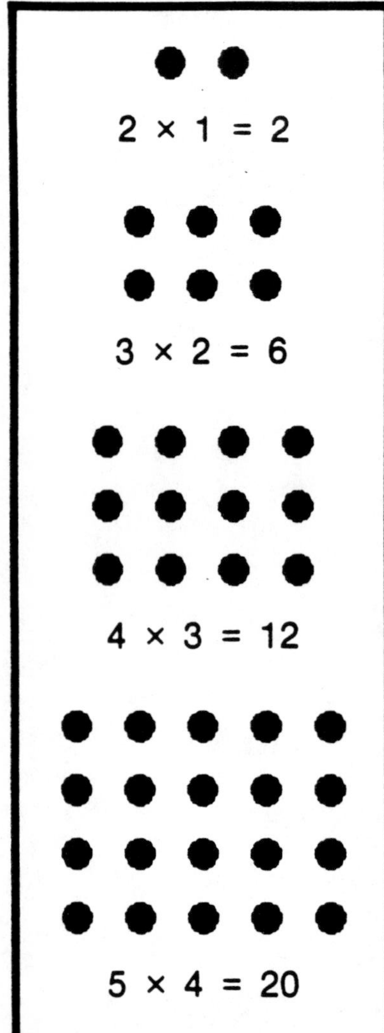

Rectangular numbers.

3.12 RECTANGULAR NUMBERS

Rectangular numbers are figurate numbers that can be represented by a rectangular array of dots. The numbers 2, 6, 12, 20, . . . are rectangular numbers. They are the number of dots used in making a successive rectangular array of dots in which the number of rows and columns increase by one.

The following program computes the first 30 rectangular numbers.

```
program RectangularNumbers;
var
  N,I        :Integer;
begin
  writeln('RECTANGULAR  NUMBERS');
  writeln('------------------------------');
  for I := 1 to 30 do
    begin
      N := I * (I + 1);
      writeln(N);
    end;
end.
```

COMPILE/RUN

```
RECTANGULAR  NUMBERS
------------------------------
2
6
12
20
30
42
56
72
90
110
132
156
182
210
240
272
306
342
380
420
462
506
552
600
650
702
756
812
870
930
```

NUMBER THEORY TRIVIA

Did you know that the sum of consecutive odd numbers gives a square number?

$$1 + 3 = 4 \quad (2^2)$$
$$1 + 3 + 5 = 9 \quad (3^2)$$
$$1 + 3 + 5 + 7 = 16 \quad (4^2)$$

In fact, the sum of the first n odd numbers is equal to n^2.

4
9
16
25
36
49
64
81
100
121
144
169
196
225
256
289
324
361
400
441
484
529
576
625
676
729
784
841
900
961

The first thirty square numbers.

3.13 SQUARE NUMBERS

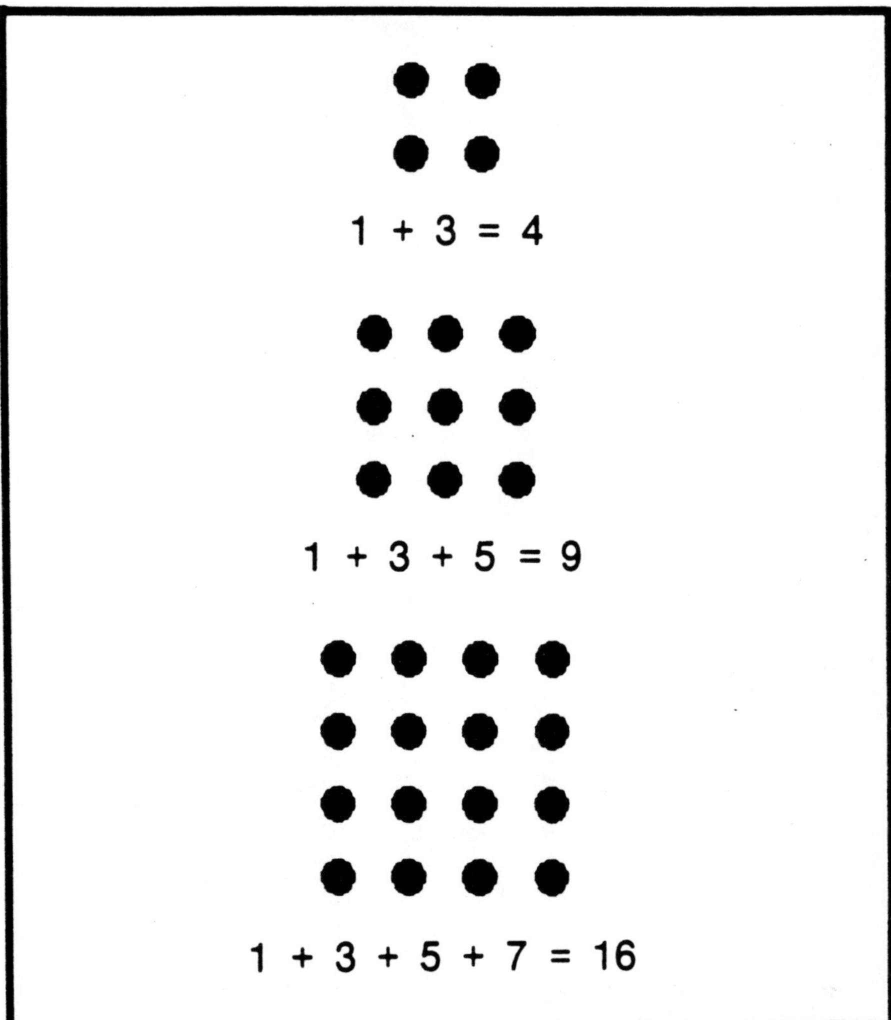

Numbers like 4, 9, 16, 25, 36, etc. are called square numbers because

$$2^2 = 4$$
$$3^2 = 9$$
$$4^2 = 16$$
$$5^2 = 25$$
$$6^2 = 36$$

and so on.

Certain pairs of numbers when added or subtracted give a square number. For example: 8 and 17

8 + 17 = 25 (a square number)
17 - 8 = 9 (a square number)

The following program finds all the pairs of numbers less than 100 that give a square number when added and when subtracted.

```
program SquareNumbers;
var
  N,P        :Integer;
begin
  writeln('N        P        N+P        P-N');
  writeln('--------------------------------');
  for N := 1 to 100 do
    for P := N+1 to 100 do
      if(SQRT(N+P) = INT(SQRT(N+P))) and (SQRT(P-N) = INT(SQRT(P-N)))
        then writeln(N:2,' ',P:2,' ',N+P:3,' ',P-N:3);
end.
```

COMPILE/RUN

N	P	N+P	P-N
4	5	9	1
6	10	16	4
8	17	25	9
10	26	36	16
12	13	25	1
12	37	49	25
14	50	64	36
16	20	36	4
16	65	81	49
18	82	100	64
20	29	49	9
24	25	49	1
24	40	64	16
28	53	81	25
30	34	64	4
32	68	100	36
36	45	81	9
36	85	121	49
40	41	81	1
42	58	100	16
48	52	100	4
48	73	121	25
54	90	144	36
56	65	121	9
60	61	121	1
64	80	144	16
70	74	144	4
72	97	169	25
80	89	169	9
84	85	169	1
96	100	196	4

NUMBER THEORY TRIVIA

The triangular numbers have the form x(x + 1)/2. A **palindromic triangular number** is one that reads the same forwards as it does backwards. The following table lists several palindromic triangular numbers.

x	x(x + 1)/2
1	1
2	3
3	6
10	55
11	66
18	171
34	595
36	666
77	3003
109	5995
132	8778
173	15051
363	66066
1111	617716
1287	828828
1593	1269261
1833	1680681
2662	3544453
3185	5073705
3369	5676765
3548	6295926
8382	35133153
11088	61477416
18906	178727871
50281	1264114621
57166	1634004361
102849	5289009825
111111	6172882716
167053	13953435931
179153	16048884061
246642	30416261403
337650	57003930075
342270	58574547585
365436	66771917766
417972	87350505378
1620621	1313207023131
3240425	5250178710525
3456734	5977618167795
3707883	6874200024786
6307938	19895044059891

3.14 STAR NUMBERS

Star numbers are numbers that can be arranged in the form of a four-pointed star. Examples are 1, 8, 21, 40, ... From the star pattern it can be shown that, except for the first star number, each star number is the sum of a square number and a triangular number.

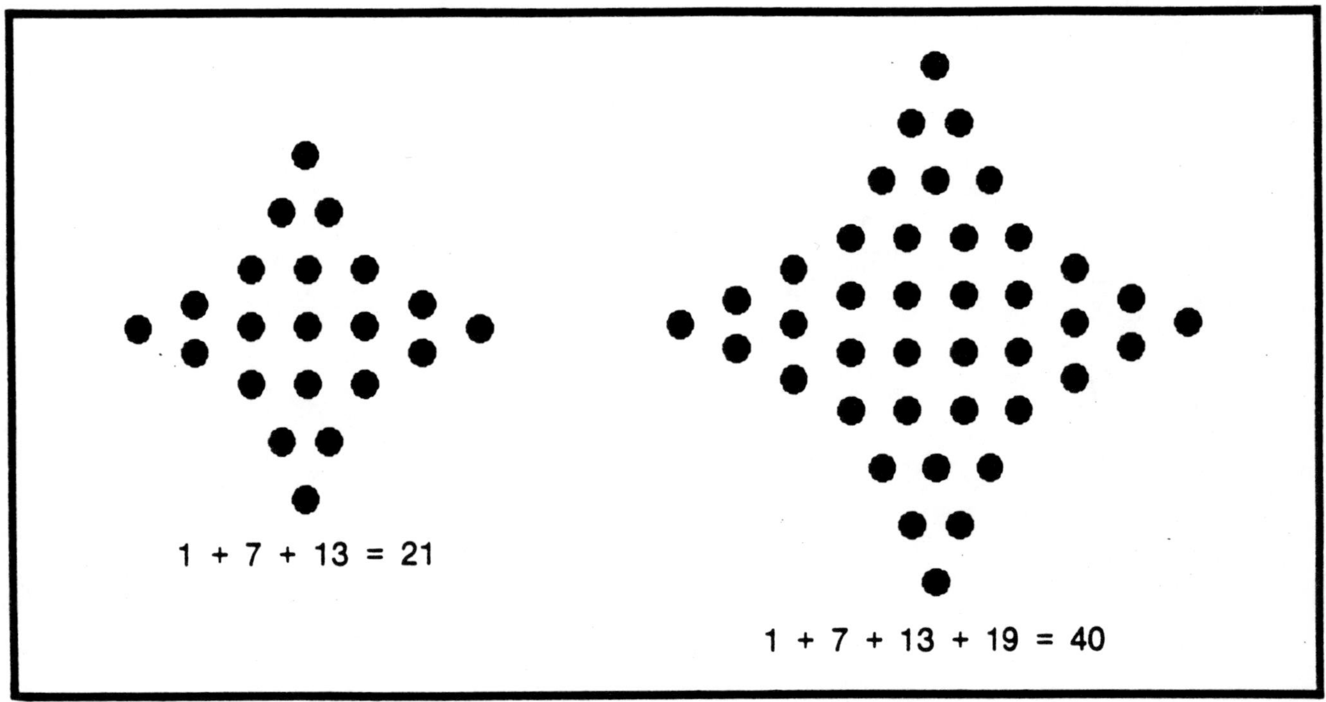

The following program determines the first 30 star numbers.

```
program StarNumbers;
var
  S,I        :Integer;
begin
  Writeln('STAR  NUMBERS');
  writeln('-------------------');
  for I := 1 to 30 do
    begin
      S := I * (3 * I - 2);
      writeln(S);
    end;
end.
```

```
STAR  NUMBERS
-------------------

1        176      645      1408
8        225      736      1541
21       280      833      1680
40       341      936      1825
65       408      1045     1976
96       481      1160     2133      2465
133      560      1281     2296      2640
```

×	1	2	3	4	5	6	7
1	1						
2		4					
3			9				
4				16			
5					25		
6						36	
7							49

The numbers on the diagonal of a multiplication square are alternately odd and even. The square of an even number is even. The square of an odd number is odd.

3.15 POWERFUL NUMBERS

Powerful numbers are integers that can be written as a sum of positive integral powers of their digits. For example, 24, 43, 63, and 89 are powerful numbers.

$$24 = 2^3 + 4^2$$
$$43 = 4^2 + 3^3$$
$$63 = 6^2 + 3^3$$
$$89 = 8^1 + 9^2$$

The following program finds integers less than 100 which can be written as the sum of their powers. (In order to conserve computer time the power was limited to 3. You may wish to increase the power in the **for** statements to 6 or 9.)

```
program PowerfulNumbers;
var
  N       :Integer;(* Counter *)
  A       :Integer;(* First digit *)
  B       :Integer;(* Second digit *)
  C       :Integer;(* Power of first digit *)
  D       :Integer;(* Power of second digit *)
function POWER (X:Real; K:Integer)      :Real;
(**** Function raises a real number X to an integer power K ****)
var
  Result     :Real;
  Index      :Integer;
begin
  Result := 1;
  for Index := 1 to K do
    Result := Result * X;
  POWER := Result;
end; (* POWER Function *)
begin
  writeln('POWERFUL NUMBERS');
  writeln('------------------------');
  for N := 1 to 99 do
    for A := 1 to 9 do
      for B := 1 to 9 do
        for C := 1 to 3 do
          for D := 1 to 3 do
            if POWER(INT(A),C)+POWER(INT(B),D=INT(N) then
              if(A = N div 10) and (B = N mod 10) then
                writeln(N,' = ',A,' ∧ ',C,' + ',B,' ∧ ',D);
end.
```

NUMBER THEORY TRIVIA

Pentagonal numbers have the form

$$P(n) = n(3n - 1)/2$$

The following table shows pentagonal numbers that are permutations of consecutive digits.

n	$P(n)$
3	12
12	210
127	24130
170	43265
465	324105
831	1035426
1470	3240615
1686	4263051

Can you write a program to compute additional pentagonal numbers of this form?

$24 = 2^3 + 4^2$
$43 = 4^2 + 3^3$
$63 = 6^2 + 3^3$
$89 = 8^1 + 9^2$
$132 = 1^1 + 3^1 + 2^7$
$135 = 1^1 + 3^2 + 5^3$
$153 = 1^1 + 5^3 + 3^3$
$175 = 1^1 + 7^2 + 5^3$
$209 = 2^7 + 0^0 + 9^2$
$224 = 2^5 + 2^7 + 4^3$
$226 = 2^1 + 2^3 + 6^3$
$262 = 2^7 + 6^1 + 2^7$
$264 = 2^5 + 6^3 + 4^2$
$264 = 2^1 + 6^1 + 4^4$
$267 = 2^1 + 6^3 + 7^2$
$283 = 2^5 + 8^1 + 3^5$
$332 = 3^4 + 3^5 + 2^3$
$333 = 3^2 + 3^4 + 3^5$
$334 = 3^3 + 3^5 + 4^3$
$357 = 3^2 + 5^1 + 7^3$
$370 = 3^3 + 7^3 + 0^0$
$371 = 3^3 + 7^3 + 1^1$
$372 = 3^3 + 7^3 + 2^1$
$373 = 3^3 + 7^3 + 3^1$
$373 = 3^4 + 7^2 + 3^5$
$374 = 3^3 + 7^3 + 4^1$
$375 = 3^3 + 7^3 + 5^1$
$375 = 3^5 + 7^1 + 5^3$
$376 = 3^3 + 7^3 + 6^1$
$377 = 3^3 + 7^3 + 7^1$
$378 = 3^3 + 7^3 + 8^1$
$379 = 3^3 + 7^3 + 9^1$
$407 = 4^3 + 0^0 + 7^3$
$445 = 4^3 + 4^4 + 5^3$
$463 = 4^1 + 6^3 + 3^5$
$518 = 5^1 + 1^1 + 8^3$
$598 = 5^1 + 9^2 + 8^3$
$629 = 6^2 + 2^9 + 9^2$
$739 = 7^1 + 3^1 + 9^3$
$794 = 7^2 + 9^3 + 4^2$
$849 = 8^3 + 4^4 + 9^2$
$935 = 9^2 + 3^6 + 5^3$
$935 = 9^3 + 3^4 + 5^3$
$994 = 9^3 + 9^1 + 4^4$
$1034 = 1^1 + 0^0 + 3^2 + 4^5$
$1073 = 1^1 + 0^0 + 7^3 + 3^6$
$1074 = 1^1 + 0^0 + 7^2 + 4^5$
$1234 = 1^1 + 2^7 + 3^4 + 4^5$
$1255 = 1^1 + 2^2 + 5^4 + 5^4$
$1306 = 1^1 + 3^2 + 0^0 + 6^4$
$1323 = 1^1 + 3^4 + 2^9 + 3^6$
$1326 = 1^1 + 3^3 + 2^1 + 6^4$
$1349 = 1^1 + 3^5 + 4^5 + 9^2$
$1364 = 1^1 + 3^1 + 6^4 + 4^3$
$1386 = 1^1 + 3^4 + 8^1 + 6^4$
$1498 = 1^1 + 4^4 + 9^3 + 8^3$
$1542 = 1^1 + 5^1 + 4^5 + 2^9$
$1634 = 1^1 + 6^4 + 3^4 + 4^4$
$1672 = 1^1 + 6^4 + 7^3 + 2^5$
$1676 = 1^1 + 6^4 + 7^3 + 6^2$
$1765 = 1^1 + 7^3 + 6^4 + 5^3$
$1836 = 1^1 + 8^3 + 3^3 + 6^4$
$2048 = 2^9 + 0^0 + 4^5 + 8^3$
$2062 = 2^3 + 0^0 + 6^1 + 2^{11}$

Powerful numbers.

REVIEW EXERCISES

1. There are at least ____________ known perfect numbers.
2. Show that there are no perfect numbers between 6 and 28.
3. Has anyone ever found an odd perfect number?
4. Show that 683 is not a perfect number.
5. Explain why no prime number can be perfect.
6. Modify the Perfect Numbers program so that it will determine the perfect number that follows 8128.
7. What is a multiperfect number?
8. Modify the Multiperfect Numbers program so that it will find all the multiperfect numbers less than 2000.
9. What is an amicable number?
10. What is an Armstrong number?
11. How do you determine a number in the tetranacci number sequence?
12. List the first 20 numbers in the Lucas number sequence.
13. Compare the tribonacci number sequence with the tetranacci and Lucas number sequences.
14. Modify the Tribonacci Numbers program so that it will compute the first sixty tribonacci numbers.
15. Lucky numbers are similar to ____________ numbers.
16. Modify the Tetrahedral Numbers program so that it will compute the first fifty tetrahedral numbers.
17. What is an abundant number?
18. What is a deficient number?
19. What is the formula used to compute a triangular number?
20. Draw a picture of the triangular number 36.
21. How do rectangular numbers differ from triangular numbers?
22. Draw a picture of the rectangular number 56.
23. What is a square number?
24. Draw a picture of the star number 96.
25. A certain number is divisible by 13. When this number is divided by the numbers 2 through 12, there is always a remainder of 1. Write a program to determine the smallest number that fits these conditions.
26. You were walking down Broadway Street in New York City and found a piece of paper. Written on the paper was a set of clues

NUMBER THEORY TRIVIA

A **tetrahedramorphic number**, X, is an integer which terminates its associated tetrahedral number T(X) = X(X+1)(X+2)/6. For example,

X	T(X)
6	56
16	816
36	8436
41	12341
56	30856
76	76076
81	91881
96	152096

Can you write a program to compute additional tetrahedramorphic numbers?

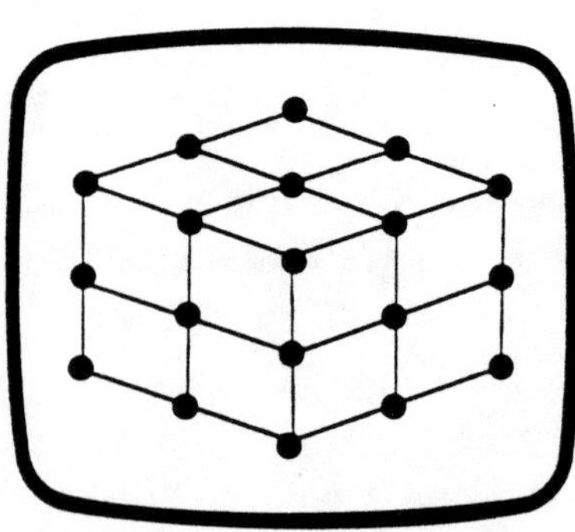

for a missing number:
When this number is divided by 10, there is a remainder of 9.
When divided by 9 there is a remainder of 8.
When divided by 7 there is a remainder of 6.
When divided by 6 there is a remainder of 5.
When divided by 5 there is a remainder of 4.
When divided by 4 there is a remainder of 3.
When divided by 3 there is a remainder of 2.
When divided by 2 there is a remainder of 1.

Write a program that will determine the smallest number that fits these clues.

27. If the sum of 1 + 2 + 3 . . . + k is a perfect square (N^2) and if N is less than 100, what are the possible values of k? For example,

    If k = 8
    $1 + 2 + 3 + 4 + 5 + 6 + 7 + 8 = 36$
    36 is a perfect square; $36 = 6^2$ and $6 < 100$.

    Write a program to find other possible values for k.

28. There are many formulas for consecutive squares whose sum is a single square.

    $1^2 + 2^2 + 3^2 + \ldots + 24^2 = 4900 = 70^2$
    $18^2 + 19^2 + 20^2 + \ldots + 28^2 = 5929 = 77^2$
    $25^2 + 26^2 + 27^2 + \ldots + 50^2 = 38025 = 195^2$
    $38^2 + 39^2 + 40^2 + \ldots + 48^2 = 20449 = 143^2$
    $456^2 + 457^2 + 458^2 + \ldots + 466^2 = 2337841 = 1529^2$
    $854^2 + 855^2 + 856^2 + \ldots + 864^2 = 8116801 = 2849^2$

    Write a program that will identify several consecutive squares whose sum equals a single square.

29. A curious relation among squares states that the sum of (n + 1) consecutive squares, beginning with the squares of n(2n + 1), is equal to the sum of the squares of the next consecutive integers.

    $3^2 + 4^2 = 5^2$
    $10^2 + 11^2 + 12^2 = 13^2 + 14^2$
    $21^2 + 22^2 + 23^2 + 24^2 = 25^2 + 26^2 + 27^2$
    $36^2 + 37^2 + 38^2 + 39^2 + 40^2 = 41^2 + 42^2 + 43^2 + 44^2$
    $55^2 + 56^2 + 57^2 + 58^2 + 59^2 + 60^2 = 61^1 + 62^2 + 63^2 + 64^2 + 65^2$

    Write a program that will illustrate this relationship.

30. Some remarkable relations exists among consecutive cubes whose sum is a cube, such as:

    $3^3 + 4^3 + 5^3 = 6^3$
    $6^3 + 7^3 + 8^3 + \ldots + 69^3 = 180^3$
    $1134^3 + 1135^3 + 1136^3 + \ldots + 2133^3 = 16830^3$

    Write a program that determines all the one-digit consecutive cubes whose sum is a cube.

4

NUMBERS WITH SPECIAL PROPERTIES

PREVIEW

For thousands of years people have played with numbers, and many of the curious problems that they have invented have come down to us. If you become really interested in numbers, you will find many things that will seem very strange. Here is a curious case relating to 45. If you take the number 987,654,321, made up of the nine digits, reverse it, and subtract, you will have three numbers – the minuend, the subtrahend, and the remainder – and the sum of the digits of each of the three is exactly 45.

Here is another curious problem: Show how to write one hundred, using only the nine figures from 1 to 9, and the signs of arithmetic. This illustration shows how it may be done.

$$100 = 1 + 2 + 3 + 4 + 5 + 6 + 7 + 8 \times 9$$

Here is another step in the endless game of numbers. You can easily see what the result of 123,456 × 8 + 6 must be, and you may care to see how far you can go in this series of numbers, and find out why you cannot go farther.

$$2^5 \times 9^2 = 2592$$

$$\begin{aligned} 1 \times 8 + 1 &= 9 \\ 12 \times 8 + 2 &= 98 \\ 123 \times 8 + 3 &= 987 \\ 1234 \times 8 + 4 &= 9876 \\ 12345 \times 8 + 5 &= 98765 \end{aligned}$$

NUMBER THEORY TRIVIA

Look at the following numbers

156 × 942 = 146952
231 × 759 = 175329
317 × 425 = 134725
437 × 602 = 263074

The three-digit numbers, each with six digits different, form a product made up of the same six digits. Can you write a program to generate all three-digit number pairs that have this property?

Another number curio is a problem like the following: find digits a, b, c, d such that

$$a^b \times c^d = abcd$$

An answer to this problem is

$$2^5 \times 9^2 = 2592$$

Some number curios appear to be occurring purely by chance. Many others are due to peculiarities in the decimal number system and do not occur when the same numbers are transformed to a base other than 10. Still others have much significance in the field of number theory.

After you finish this chapter, you should be able to:

1. Identify wraparound numbers, multigrades, interesting identities, consecutive cubes, and many other interesting number relationships.
2. See how a microcomputer can be used to produce interesting number relationships.
3. Write programs to generate several interesting number relationships.

4.1 WRAPAROUND NUMBERS

A wraparound number is a number where each digit describes where the next one is located (by counting to the right and wrapping around when needed). All the digits are landed on, once

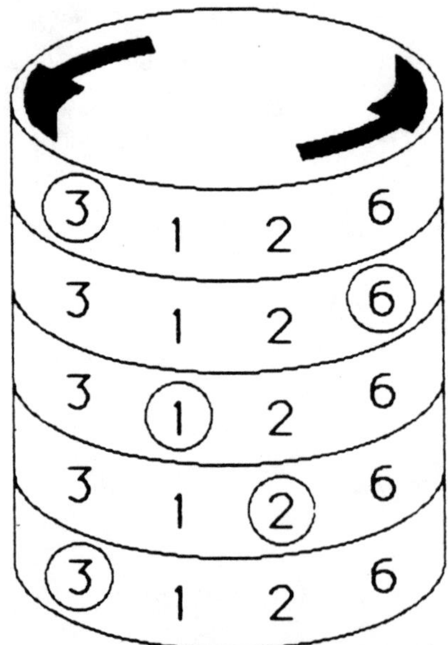

each. After using each digit once, you arrive back at the original (leftmost) digit. For example, 3126 is a wraparound number. Start with the leftmost digit, the 3. Take the number 3 and count three digits to the right landing on 6. Take the number 6 and count six digits to the right, wrapping around to the left, landing on 1. Take

NUMBER THEORY TRIVIA

The set of prime numbers is denumerable; that is the prime numbers can be placed in one-to-one correspondence with the natural numbers. The *n*th prime is denoted *n*. For example,

$P_1 = 2$
$P_2 = 3$
$P_5 = 11$
$P_9 = 23$
$P_{21} = 73$

NUMBER THEORY TRIVIA

There is a series of integers which always converge to 1, given by:

If n be even, the next $n = n/2$
If n be odd, the next $n = 3n+1$

Start with any positive integer, n. Divide it by 2. If it be even (no remainder) the quotient is the next number. If odd (remainder of 1) multiply by 3 and add 1. Then repeat for each number you obtain. Each series will move to 1 without repeating any number. Shown below are the series of the first ten positive integers.

n	1	2	3	4	5	6	7	8	9	10
	4	1	10	2	16	3	22	4	28	5
	2		5	1	8	10	11	2	14	16
	1		16		4	5	34	1	7	8
			8		2	16	17		22	4
			4		1	8	52		11	2
			2			4	26		34	1
			1			2	13		17	
						1	40		52	
							20		26	
							10		13	
							5		40	
							16		20	
							8		10	
							4		5	
							2		16	
							1		8	
									4	
									2	
									1	

the number 1 and count one digit to the right, landing on 2. Take the number 2 and count two digits to the right, wrapping around when needed, landing on 3.

The following program finds all four-digit numbers starting with the digit 3 which are wraparound numbers.

```
program WraparoundNumbers;
label 10;
var
  A          :array[1 . . 4] of Integer;
  I,K,N      :Integer;
  R,S        :Integer;
begin
  (**** Print heading ****)
  writeln('WRAPAROUND NUMBERS');
  writeln('----------------------------');
  writeln;
  for N := 3000 to 3999 do
    begin
      R := N;
      I := 4;
      repeat
        S := R div 10;
        A[I] := R - 10 * S;
        R := S;
        I := I - 1;
      until I < = 0;
      K := 1;
      for I := 1 to 4 do
        begin
          if A[K] = 0 then goto 10;
          S := K + A[K];
          A[K] := 0;
          K := S - 4 * ((S - 1) div 4);
        end;
      if K = 1 then writeln(N);
      10:(* Look for new wraparound number *)
    end;
end.
```

COMPILE/RUN

```
WRAPAROUND NUMBERS
----------------------------

3122
3126
3162
3166
3333
3337
3373
3377
3522
3526
3562
3566
3733
3737
3773
3777
3922
3926
3962
3966
```

4.2 INTEGER ABCD

The following program finds four integers A, B, C, and D which satisfy

$$A^B \times C^D = ABCD$$

```
program IntegerABCD;
var
  A,B,C,D      :Integer;
function POWER (X:Real; K:Integer) :Real;
(**** Function raises a real number X to an integer power K ****)
var
  Result      :Real;
  Index       :Integer;
begin
  Result := 1;
  for Index := 1 to K do
    Result := Result * X;
  POWER := Result;
end; (* POWER Function *)
begin
  writeln('A        B        C        D');
  writeln('-------------------------');
  for A := 1 to 9 do
    for B := 1 to 9 do
      for C := 1 to 9 do
        for D := 1 to 9 do
          if POWER(A,B) * POWER(C,D) = 1000*A + 100*B + 10*C + D then
            writeln(A, B:4, C:4, D:4);
end.
```

COMPILE/RUN

```
A    B    C    D
----------------------------
2    5    9    2
```

4.3 NON-REPEATING DIGITS

During a lesson on combinations, Tom's teacher stated that there are exactly 648 three-digit integers with no repeated digits. While the teacher rambled on, Tom played around with the number 648. He added its digits, squared the sum, and divided that sum into 648. He was pleased to find no remainder.

Of all the other three-digit numbers with no repeating digits, how many have no remainder when divided by the square of the sum of the digits? What are they? The following program finds these numbers.

```
program NonRepeatingDigits;
var
  T         :Integer;(* Numbers with no remainder *)
  N         :Integer;(* Number *)
  S         :Integer;(* Square of digit sum *)
  A,B,C     :Integer;(* Counters and digits of number *)
begin
  writeln('NUMBER        SQUARE OF        NUMBER/SQUARE');
  writeln('              SUM OF DIGITS');
  writeln('-----------------------------------------------------------');
  T := 0;
  for A := 1 to 9 do
    for B := 0 to 9 do
      for C := 0 to 9 do
        if(A-B)*(B-C)*(A-C) <> 0 then
          begin
            (**** Add digits of number ****)
            N := 100 * A + 10 * B + C;
            (**** Square sum of digits ****)
            S := (A+B+C) * (A+B+C);
            (**** Check for remainder ****)
            if N/S - TRUNC(N/S) = 0 then
              begin
                T := T + 1;
                writeln(N:4,'          ',S:5,'          ',N div S);
              end;
          end;
  writeln;
  writeln('NUMBERS WITH NO REMAINDER = ',T);
end.
```

COMPILE/RUN

```
NUMBER        SQUARE OF      NUMBER/SQUARE
             SUM OF DIGITS
-------------------------------------------------------
  162            81              2
  243            81              3
  324            81              4
  392           196              2
  405            81              5
  512            64              8
  605           121              5
  648           324              2
  810            81             10
  972           324              3

NUMBERS WITH NO REMAINDER = 10
```

NUMBER THEORY TRIVIA

G.H. Hardy stated that 8712 and 9801 are the only four figure numbers which are multiples of their reversals,

8712 = 4 × 2178

9801 = 9 × 1089

The following table lists all numbers less than 10^9 having the reversible multiply property.

```
 8712 = 4 × 2178
 9801 = 9 × 1089
 87912 = 4 × 21978
 98901 = 9 × 10989
 879912 = 4 × 219978
 989901 = 9 × 109989
 5604390 = 6 × 934065
 879912 = 4 × 2199978
 9899901 = 9 × 1099989
 56094390 = 6 × 9349065
 87128712 = 4 × 21782178
 87999912 = 4 × 21999778
 98019801 = 9 × 10891089
 560994390 = 6 × 93499065
 871208712 = 4 × 217802178
 879999912 = 4 × 219999978
 980109801 = 9 × 108901089
 989999901 = 9 × 109999989
```

4.4 INTEGER SUMS

Did you realize that the square of the sum of N integers equals the sum of the cubes of each individual integer. For example,

1 2 3 4 = 10 $10^2 = 100$

$1^3 + 2^3 + 3^3 + 4^3 = 100$

The following program verifies this property for values of N equal to 1, 2, 3, . . ., 12.

```
program IntegerSums;
var
  N      :Integer;(* Number *)
  X      :Integer;(* Sum of squares *)
  Y      :Integer;(* Sum of cubes *)
begin
  X := 0;
  Y := 0;
  writeln('VALUE          SUM OF          SUM OF');
  writeln('OF N           SQUARES          CUBES ');
  writeln('-----------------------------------------------');
  for N := 1 to 12 do
    begin
      X := X + N;
      Y := Y + N * N * N;
      writeln(N:4,X*X:10,Y:12);
    end;
end.
```

COMPILE/RUN

VALUE OF N	SUM OF SQUARES	SUM OF CUBES
1	1	1
2	9	9
3	36	36
4	100	100
5	225	225
6	441	441
7	784	784
8	1296	1296
9	2025	2025
10	3025	3025
11	4356	4356
12	6084	6084

NUMBER THEORY TRIVIA

Some large prime numbers

107,928,278,317
11,000,001,446,613,353
12,345,678,901,234,567,891
357,686,312,646,216,567,629,137

4.5 SUM AND PRODUCT

Some numbers have the property that the sum and product of the digits are equal. For example, 22 has the property because 2 + 2 = 2 × 2, but 67 does not because 6 + 7 does not equal 6 × 7. To see if any three-digit numbers have the property, we must test every number of the form abc to see whether a + b + c = a × b × c. The following program accomplishes this task.

```
program SumAndProduct;
var
  I,J,K,Sum,Product        :Integer;
begin
  writeln('THE SUM AND PRODUCT OF THE DIGITS ARE EQUAL');
  writeln('---------------------------------------------------------');
  for I := 1 to 9 do
    for J := I to 9 do
      for K := J to 9 do
        begin
          Sum := I + J + K;
          Product := I * J * K;
          (**** Does the Sum equal the Product? ****)
          if Sum = Product then
            writeln('THE DIGITS ARE ', I, ' , ', J, ' AND ', K);
        end;
end.
```

COMPILE/RUN

```
SUM AND PRODUCTS OF THE DIGITS ARE EQUAL
---------------------------------------------------------
THE DIGITS ARE 1, 2, AND 3
```

In the **for** statements, zero is excluded as a possible digit because if any digit were zero, the product of the digits would be zero, and therefore the product could not equal the sum. The analysis produces just one set of digits: 1, 2, and 3. All possible permutations of these digits produce six solutions to the problem: 123, 132, 213, 231, 312, and 321. If the reader wishes to have the program print out all possible permutations, change two of the **for** statements to read

```
for J := 1 to 9 do
```

and

```
for K := 1 to 9 do
```

NUMBER THEORY TRIVIA

Catalan numbers are defined by the formula

$$C_n = \frac{(2n)!}{n!(n+1)!}$$

where n = 0, 1, 2, . . . The first few Catalan numbers are

1,1,2,5,14,42,132,429,1430,4862,. . .

A few other large Catalan numbers are

C_{16} = 35,357,670
C_{24} = 1,289,904,147,324
C_{61} = 6,182,127,958,584,855,650 487,080,847,216,336
C_{64} = 368,479,169,875,816,659, 479,009,042,713,546,950

4.6 UNUSUAL MULTIPLICATION

Michael walked into his mathematical analysis class one day and saw this problem on the chalkboard:

2178	x	4	=	8712
multiplicand		multiplier		product

The class was studying multiplication problems involving four-digit multiplicands and one-digit multipliers. Michael noticed the digits in the multiplicand, 2178, reversed gave the correct product 8712. He asked the teacher if any other four-digit by one-digit multiplication problems had the same unusual property. The teacher replied that there are several obvious examples for the multiplier 1 (such as 2222 × 1 = 2222) but only one other example when the single-digit multiplier was greater than one.

The following program finds the multiplier and the corresponding four-digit multiplicand.

```
program UnusualMultiplication;
var
  D,N,P,R        :Integer;
  K              :Real;
begin
  writeln('MULTIPLICAND        MULTIPLIER        PRODUCT');
  writeln('----------------------------------------------------');
  for N := 1000 to 9999 do
    begin
      R := 0;
      P := N;
      repeat
        K := P/10;
        P := TRUNC(K);
        D := TRUNC(10 * (K - P) + 0.5);
        R := 10 * R + D;
      until P = 0;
      if R/N = TRUNC(R/N) then
        if R <> N then
          if N <> 2178 then
            begin
              writeln(N:8,'            ',R div N,'            ',R);
              HALT;
            end;
    end;
end.
```

COMPILE/RUN

```
MULTIPLICAND        MULTIPLIER        PRODUCT
------------------------------------------------
    1089                9              9801
```

NUMBER THEORY TRIVIA

The number 2 can be written as $1^2 + 1^2$. The number 50 is the smallest integer which can be written in *two* different ways as the sum of the squares of two integers:

$$50 = 1^2+7^2 = 5^2+5^2$$

The number 325 is the smallest integer which can be written in *three* different ways as the sum of the squares of two integers:

$$325 = 1^2+18^2 = 6^2+17^2 = 10^2+15^2$$

Can you write a program to find other smallest integers X(*n*) which can be written in exactly *n* different ways as the sum of the squares of two integers?

NUMBER THEORY TRIVIA

Every pandigital integer (a nine-digit integer that contains each of the digits 1 through 9, or a ten-digit integer that contains each of the digits 0 through 9) can be expressed as the sum of one or more factorials. For example,

479365218 = 3!+3!+3!+ 6!+ 9!+12!
518923476 = 3!+3!+4!+ 7!+11!+12!
958371246 = 3!+5!+7!+ 9!+12!+12!
6231057984 = 4!+5!+7!+ 8!+ 9!+10!+13!
7185392640 = 6!+7!+9!+12!+12!+13!
7185064392 = 4!+4!+4!+ 8!+12!+12!+13!

Can you write a program to express other pandigital integers as the sum of factorials.

4.7 SPECIAL FOUR-DIGIT NUMBERS

The four-digit number 3025 is special. The sum of the first two digits (30) and the last two digits (25) is 55. If you square 55 the original number is obtained ($55^2 = 3025$).

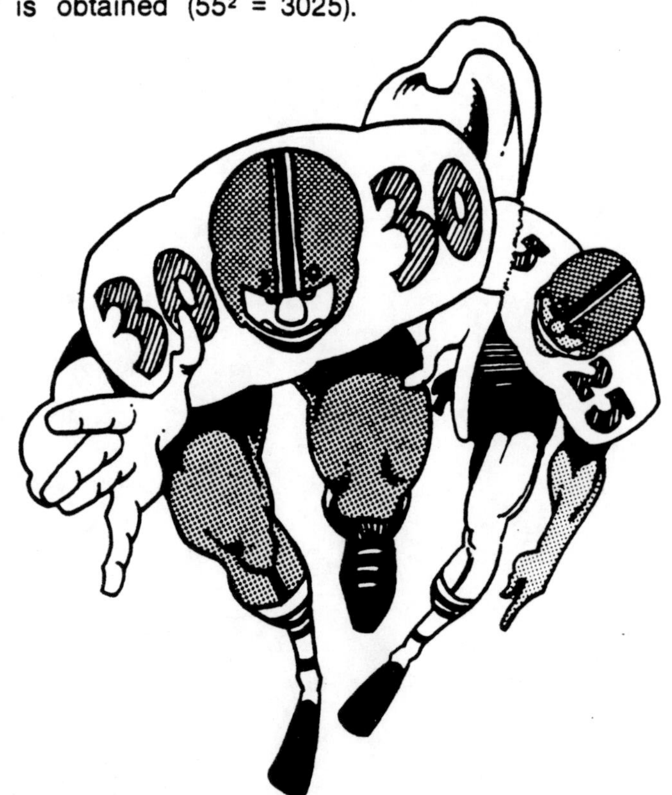

The following program determines all four-digit numbers that have this property.

```
program SpecialFourDigitNumbers;
var
  N,F,L        :Integer;
begin
  writeln('N              SUM                       SUM  SQUARED');
  writeln('------------------------------------------------------');
  for N := 1000 to 9999 do
    begin
      F := N div 100;
      L := N - 100 * F;
      if (F + L) * (F + L) = N then
        writeln(N, '        ',F+L, '                ',(F+L) * (F+L));
    end;
end.
```

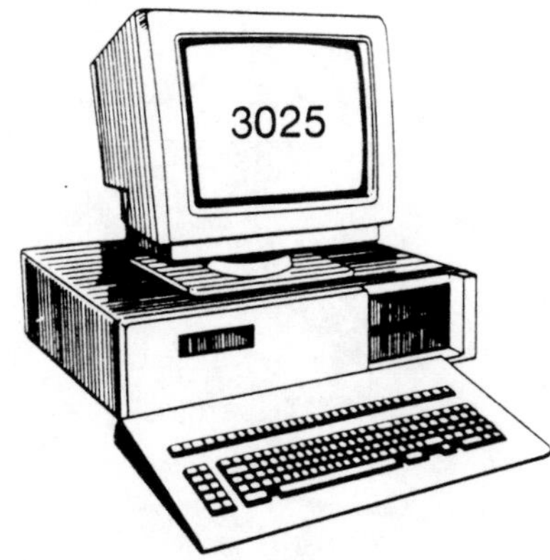

COMPILE/RUN

```
N            SUM              SUM  SQUARED
------------------------------------------
2025          45               2025
3025          55               3025
9801          99               9801
```

4.8 FOUR INTEGERS

The following program finds four integers W, X, Y, and Z which satisfy

$$W^3 + X^3 + Y^3 = Z^3$$

```
program FourIntegers;
var
  W,X,Y,Z       :Integer;
begin
  writeln(' W  X  Y        Z ');
  writeln('----------------------');
  for W := 1 to 9 do
    for X := 1 to 9 do
      for Y := 1 to 9 do
        for Z := 1 to 9 do
          if W*W*W + X*X*X + Y*Y*Y = Z*Z*Z then
            writeln(W:2, X:4, Y:4, X:7);
end.
```

COMPILE/RUN

```
W  X  Y     Z
--------------------------
1  6  8     9
```

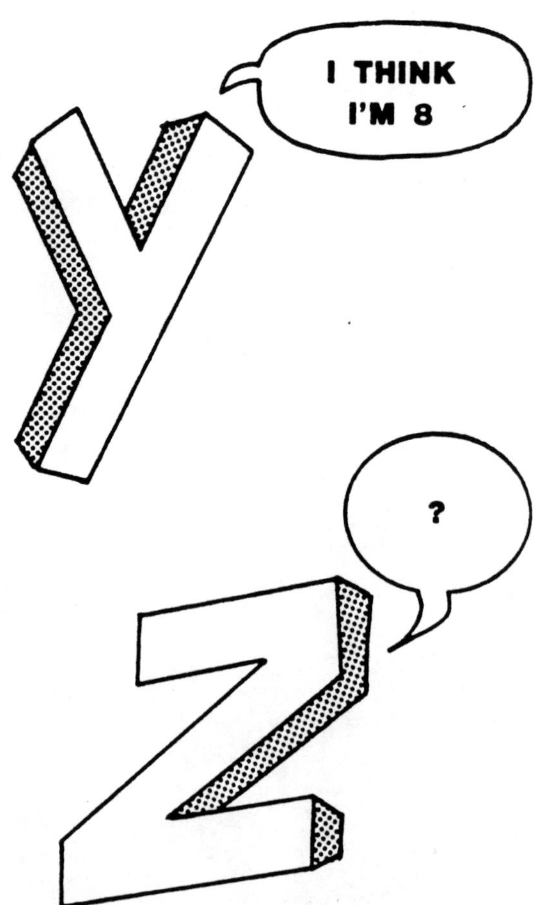

4.9 SUM OF TWO SQUARES

The following program finds all integers less than 50 which can be written as the sum of two squares. For example

$$34 = 3^2 + 5^2$$

```
program SumOfTwoSquares;
var
  A,B,N      :Integer;
begin
  for N := 1 to 50 do
    for A := 1 to 7 do
      for B := 1 to 7 do
        if A*A + B*B = N then
          writeln(N, ' = ', A*A:2, ' + ', B*B:2);
end.
```

COMPILE/RUN

```
2 = 1 + 1
5 = 1 + 4
8 = 4  + 4
10 = 1 + 9
13 = 4 + 9
17 = 1 + 16
18 = 9 + 9
20 = 4 + 16
25 = 9 + 16
26 = 1 + 25
29 = 4 + 25
32 = 16 + 16
34 = 9 + 25
37 = 1 + 36
40 = 4 + 36
41 = 16 + 25
45 = 9 + 36
50 = 1 + 49
```

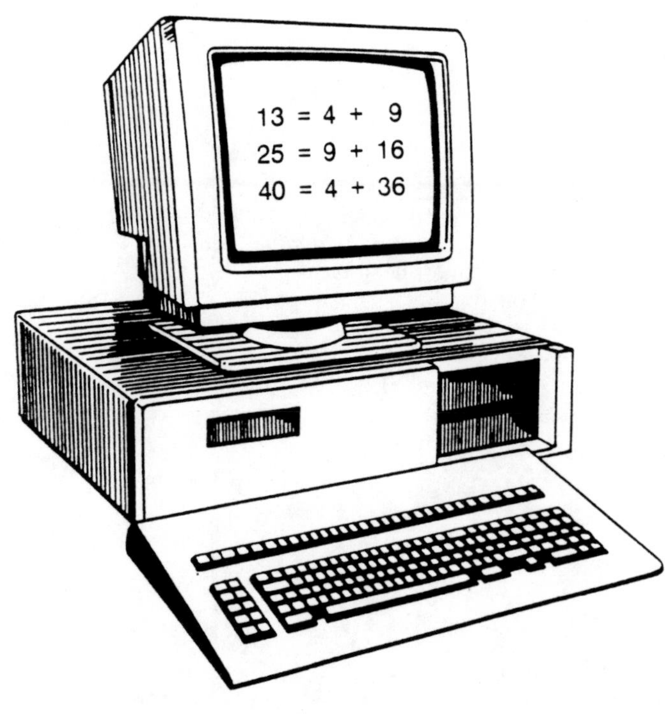

NUMBER THEORY TRIVIA

23 is a **happy number** because

23 → 2 × 2 + 3 × 3 = 13
13 → 1 × 1 + 3 × 3 = 10
10 → 1 × 1 + 0 × 0 = 1

Since the sequence of numbers 23, 13, 10, and 1 ends with the number 1, the number 23 is said to be a happy number. A few happy numbers are 10, 13, 19, 23, 28, 31, 32, 44, . . .

Can you write a program to generate all happy numbers less than 100?

NUMBER THEORY TRIVIA

The set of positive integers which are divisible by the sum of their digits are called **Niven numbers** (after Ivan Niven). N is called a Niven number if S(N) is a factor of N, where S(N) is the sum of the digits of N. There is an infinite number of Niven numbers. The first few are 1, 2, 3, 4, 5, 6, 7, 8, 9, 10, 12, 18, 20, 21, 24, . . .

Can you write a program to compute Niven numbers?

4.10 INTERESTING IDENTITIES

Note the following:

$$3^4 + 4^4 + 5^4 = 5^2 + 19^2 + 24^2$$
$$3^8 + 4^8 + 5^8 = 5^4 + 19^4 + 24^4$$

$$7^2 + 34^2 + 41^2 = 14^2 + 29^2 + 43^2$$
$$7^4 + 34^4 + 41^4 = 14^4 + 29^4 + 43^4$$

These identities are most unusual. It is very difficult (unless you use a microcomputer) to find three numbers a, b and c such that

$$a^n + b^n + c^n \text{ and } a^{2n} + b^{2n} + c^{2n}$$

equals

$$d^{n/2} + e^{n/2} + f^{n/2} \text{ and } d^n + e^n + f^n$$

respectively.

The program on page 95 finds the identities for

$$A^p + B^p + C^p = D^{p/2} + E^{p/2} + F^{p/2}$$
$$A^{2p} + B^{2p} + C^{2p} = D^p + E^p + F^p$$

for values of A, B, C, D, E, F equal to 2 through 9, and p equal to 2 through 5.

It's the most important discovery of the century! Come on, let's relax.

NUMBER THEORY TRIVIA

In 1742, Christian Goldbach, a Russian mathematician, in a letter to Leonhard Euler, postulated that every even number larger than 4 can be written as the sum of two primes. For example,

16 = 5 + 11
24 = 19 + 5
38 = 7 + 31
66 = 5 + 61
80 = 19 + 61
94 = 23 + 71
156 = 67 + 89
162 = 13 + 149
388 = 107 + 281
398 = 199 + 199
1000 = 491 + 509
1644 = 653 + 991

```
program InterestingIdentities;
var
  A,B,C,D,E,F      :Integer;(* Numbers 2 — 9 *)
  P                :Integer;(* Power 2 *)
  W,X,Y,Z          :Integer;(* Number sums *)
function POWER (X:Real; K:Integer):Real;
(**** Function raises a real number X to an integer power K ****)
var
  Result     :Real;
  Index      :Integer;
begin
  Result := 1;
  for Index := 1 to K do
    Result := Result * X;
  POWER := Result;
end; (* POWER Function *)
begin
  writeln('        A        B        C        D        E        F        P');
  writeln('        ------------------------------------------------------------');
  P := 2;
    for A := 2 to 9 do
      for B := A to 9 do
        for C := B to 9 do
          for D := C to 9 do
            for E := D to 9 do
              for F := E to 9 do
          begin
            (**** Calculate sums W, X, Y and Z ****)
            W := TRUNC(POWER(INT(A),P)+POWER(INT(B),P)+POWER(INT(C),P));
            X := TRUNC(POWER(INT(A),2*P)+POWER(INT(B),2*P)+POWER(INT(C),2*P));
            Y := TRUNC(POWER(INT(D),P div 2)+POWER(INT(E),P div 2)+POWER(INT(F),P div 2));
            Z := TRUNC(POWER(INT(D),P)+POWER(INT(E),P)+POWER(INT(F),P));
            if (W = Y) AND (X = Z) then
              writeln(A:5,B:5,C:5,D:5,E:5,F:5,P:5);
          end;
end.
```

4.11 POWERS OF DIGITS

Sums of powers of digits, equal to sums of like powers of other digits, serve as generators of powers whose sum is the same as that of these powers with the digits reversed. Thus, starting with

$$4^2 + 5^2 + 6^2 = 8^2 + 3^2 + 2^2$$

we can combine in any manner the digits on the right- and left-hand side of the equal sign and obtain, for example, 48, 53, and 62. These numbers, with the digits reversed, are 84, 35, and 26. Then

$$48^2 + 53^2 + 62^2 = 84^2 + 35^2 + 26^2$$

The reader may wish to write a program that will generate number arrangements that satisfy this relationship.

NUMBER THEORY TRIVIA

Some numbers can be expressed as the sum of the squares of two positive integers in several ways, e.g.:

$$325 = 1^2 + 18^2 = 6^2 + 17^2 = 10^2 + 15^2$$

NUMBER THEORY TRIVIA

A **trimorphic number** is one whose cube ends with itself, for example, $51^3 = 132651$ and $875^3 = 669921875$. Can you write a program to generate several trimorphic numbers?

4.12 MULTIGRADES

It is not difficult to find a number of squares whose sum is equal to the sum of other squares, for example, $6^2 + 5^2 + 4^2 = 3^2 + 2^2 + 8^2$. More noteworthy are such relations which hold simultaneously for several powers; thus

$$1^n + 6^n + 8^n = 2^n + 4^n + 9^n \qquad (n = 1, 2)$$

$$1^n + 4^n + 5^n + 5^n + 6^n + 9^n = 2^n + 3^n + 3^n + 7^n + 7^n + 8^n$$
$$(n = 1, 2, 3)$$

$$1^n + 5^n + 8^n + 12^n + 18^n + 19^n = 2^n + 3^n + 9^n + 13^n + 16^n + 20^n$$
$$(n = 1, 2, 3, 4)$$

Such relationships are called "multigrades." The symbol $\stackrel{n}{=}$ is used to indicate the range of powers for which a multigrade holds; the "trigrade" above, in abbreviated form, would be written 1, 4, 5, 5, 6, 9 $\stackrel{3}{=}$ 2, 3, 3, 7, 7, 8. This saves writing the exponent for each power. More remarkable yet are the pentagrades 0, 5, 6, 16, 17, 22 $\stackrel{5}{=}$ 1, 2, 10, 12, 20, 21; and 1, 11, 13, 33, 35, 45 $\stackrel{5}{=}$ 3, 5, 21, 25, 41, 45. If we add the respective terms of these two series we get 1, 16, 19, 49, 52, 67 $\stackrel{5}{=}$ 4, 7, 31, 37, 61, 66.

The following program determines other pentagrades. The program uses the relationships: a, (a + 4b + c), (a + b + 2c), (a + 9b + 4c), (a + 6b + 5c), (a + 10b + 6c) $\stackrel{5}{=}$ (a + b), (a + c), (a + 6b + 2c), (a + 4b + 4c), (a + 10b + 5c), (a + 9b + 6c).

```
program Pentagrades;
var
  L1,L2,L3,L4,L5,L6          :Integer;(* Left side of formula *)
  R1,R2,R3,R4,R5,R6          :Integer;(* Right side of formula *)
  A,B,C                      :Integer;(* Computational variables *)
begin
  writeln('PENTAGRADES');
  writeln('-----------------');
(**** Compute and print pentagrades ****)
  for C := 1 to 3 do
    for B := 1 to 3 do
      for A := 1 to 3 do
        begin
          L1 := A;
          L2 := A + 4 * B + C;
          L3 := A + B + 2 * C;
          L4 := A + 9 * B + 4 * C;
          L5 := A + 6 * B + 5* C;
          L6 := A + 10 * B + 6 * C;
          R1 := A + B;
          R2 := A + C;
          R3 := A + 6 * B + 2 * C;
          R4 := A + 4 * B + 4 * C;
```

```
            R5 := A + 10 * B + 5 * C;
            R6 := A + 9 * B + 6 * C;
            writeln(L1,',',L2,',',L3,',',L4,',',L5,',',L6,'(5 =)',R1,',',R2,',',R3,',',R4,',',R5,',',R6);
          end;
end.
```

```
COMPILE/RUN

PENTAGRADES
------------------
1, 6, 4, 14, 12, 17 (5 =) 2, 2, 9, 9, 16, 16
2, 7, 5, 15, 13, 18, (5 =) 3, 3, 10, 10, 17, 17
3, 8, 6, 16, 14, 19 (5 =) 4, 4, 11, 11, 18, 18
1, 10, 5, 23, 18, 27 (5 =) 3, 2, 15, 13, 26, 25
2, 11, 6, 24, 19, 28 (5 =) 4, 3, 16, 14, 27, 26
3, 12, 7, 25, 20, 29 (5 =) 5, 4, 17, 15, 28, 27
1, 14, 6, 32, 24, 37 (5 =) 4, 2, 21, 17, 36, 34
2, 15, 7, 33, 25, 38 (5 =) 5, 3, 22, 18, 37, 35
3, 16, 8, 34, 26, 39 (5 =) 6, 4, 23, 19, 38, 36
1, 7, 6, 18, 17, 23 (5 =) 2, 3, 11, 13, 21, 22
2, 8, 7, 19, 18, 24 (5 =) 3, 4, 12, 14, 22, 23
3, 9, 8, 20, 19, 25 (5 =) 4, 5, 13, 15, 23, 24
1, 11, 7, 27, 23, 33 (5 =) 3, 3, 17, 17, 31, 31
2, 12, 8, 28, 24, 34 (5 =)4, 4, 18, 18, 32, 32
3, 13, 9, 29, 25, 35 (5 =) 5, 5, 19, 19, 33, 33
1, 15, 8, 36, 29, 43 (5 =) 4, 3, 23, 21, 41, 40
2, 16, 9, 37, 30, 44 (5 =) 5, 4, 24, 22, 42, 41
3, 17, 10, 38, 31, 45 (5 =) 6, 5, 25, 23, 43, 42
1, 8, 8, 22, 22, 29 (5 =) 2, 4, 13, 17, 26, 28
2, 9, 9, 23, 23, 30 (5 =) 3, 5, 14, 18, 27, 29
3, 10, 10, 24, 24, 31 (5 =) 4, 6, 15, 19, 28, 30
1, 12, 9, 31, 28, 39 (5 =) 3, 4, 19, 21, 36, 37
2, 13, 10, 32, 29, 40 (5 =) 4, 5, 20, 22, 37, 38
3, 14, 11, 33, 30, 41 (5 =) 5, 6, 21, 23, 38, 39
1, 16, 10, 40, 34, 49 (5 =) 4, 4, 25, 25, 46, 46
2, 17, 11, 41, 35, 50 (5 =) 5, 5, 26, 26, 47, 47
3, 18, 12, 42, 36, 51 (5 =) 6, 6, 27, 27, 48, 48
```

NUMBER THEORY TRIVIA

Call a ten-digit number **pandigital** if it is a permutation of the digits 0 1 2 3 4 5 6 7 8 9. The smallest pandigital number x such that 2x, 4x, and 8x are also pandigital is 0123456789 and the largest is 1234567890.

Can you write a program to determine the next smallest and next largest such numbers?

4.13 PECULIAR PROPERTY

The number 48 has the peculiar property that if one is added to it, the sum is a square, 49, and if one is added to its half, 24, the result, 25, is also a square.

NUMBER THEORY TRIVIA

An **automorphic number** is an integer whose square ends with the given integer as

$$6^2 = 36$$
$$25^2 = 625$$
$$76^2 = 5776$$
$$625^2 = 390625$$

Can you write a program to generate several automorphic numbers?

The following program finds that 48 and 1680 also have this property. The reader can generate additional numbers by modifying the **for** statement. Other numbers are

57120
1940448
65918160
2239277040
76069501248
2584123765440

```
program PeculiarProperty;
var
  N       :Integer;(* Number and looping counter *)
begin
  writeln('            N      N+1      N/2+1');
  writeln('            ---    ---        --- ');
  (**** Compute numbers less than 2000 ****);
  for N := 1 to 2000 do
    begin
      if INT(SQRT(N+1)) = SQRT(N+1) then
        if INT(SQRT(N/2+1)) = SQRT(N/2+1) then
          writeln(N:8,(N+1):7,(N div 2+1):8);
    end;
end.
```

COMPILE/RUN

```
      N      N+1    N/2+1
    ---      ---      ---
     48       49       25
   1680     1681      841
```

4.14 CONSECUTIVE CUBES

The cubes 1; 8; 27; 64; 125; . . . are related to triangular numbers. The sum of N consecutive cubes starting from unity is equal to the square of the Nth triangular number. For N = 4, we have

$$1^3 + 2^3 + 3^3 + 4^3 = 100$$

which is the square of the fourth triangular number 10. In general,

$$1^3 + 2^3 + 3^3 + . . . + N^3 = (N(N + 1) / 2)^2$$

where N(N + 1) / 2 is the Nth triangular number.

The following program will demonstrate this relationship.

```
program ConsecutiveCubes;
var
  N,T,C,L       :Integer;
begin
  write('ENTER NUMBER OF TERMS: ');
  readln(L);
  writeln;
  writeln('TERM      SUM OF CUBES      SQUARE OF TRIANGULAR NO.');
  writeln('-------------------------------------------------------------');
  C := 0;
  for N := 1 to L do
    begin
      (**** Compute triangular number ****)
      T := (N * (N + 1)) div 2;
      (**** Compute sum of consecutive cubes ****)
      C := C + N * N * N;
      (**** Print term, sum of cubes and square of triangular number ****)
      writeln(N:2,'          ', C:6, '                    ', T*T:6);
    end;
end.
```

COMPILE/RUN

```
ENTER NUMBER OF TERMS: 25

TERM      SUM OF CUBES      SQUARE OF TRIANGULAR NO.
-------------------------------------------------------------
 1              1                    1
 2              9                    9
 3             36                   36
 4            100                  100
 5            225                  225
 6            441                  441
 7            784                  784
 8           1296                 1296
 9           2025                 2025
10           3025                 3025
11           4356                 4356
12           6084                 6084
13           8281                 8281
14          11025                11025
15          14400                14400
16          18496                18496
17          23409                23409
18          29241                29241
19          36100                36100
20          44100                44100
21          53361                53361
22          64009                64009
23          76176                76176
24          90000                90000
25         105625               105625
```

I DO NOT
LIKE TO WORK
WITH LONELY
NUMBERS

NUMBER THEORY TRIVIA

A **Smith number** (as defined by A. Wilansky) is a composite number, the sum of whose digits is equal to the sum of the digits of a prime factorization. For example,

$$4937775 = 3 \times 5 \times 5 \times 65837.$$

The question of whether there are infinitely many Smith numbers or not remains open.

4.15 SUM OF THREE CUBES

Each number is equal to the sum of the cubes of its 3 sections. Thus,

$$34 \quad 10 \quad 67 = 341067 = 34^3 + 10^3 + 67^3$$

The following program determines all six-digit numbers with this property.

```
program SumOfThreeCubes;
var
  A      :Integer;(* Leftmost part of number *)
  B      :Integer;(* Center part of number *)
  C      :Integer;(* Rightmost part of number *)
  X      :Integer;(* Calculation variable *)
begin
  writeln('NUMBER          A         B          C');
  writeln('----------      -----------------------');
  for A := 10 to 99 do
    for B := 0 to 99 do
      for C := 0 to 99 do
        begin
          X := A * 10000 + B * 100 + C;
          if A*A*A + B*B*B + C*C*C = X then
            writeln(X:7,'          ',A:2,'          ',B:2,'          ',C:2);
        end;
end.
```

COMPILE/RUN

NUMBER	A	B	C
165033	16	50	33
221859	22	18	59
336700	33	67	0
336701	33	67	1
340067	34	0	67
341067	34	10	67
407000	40	70	0
407001	40	70	1
444664	44	46	64
487215	48	72	15
982827	98	28	27
983221	98	32	21

NUMBER THEORY TRIVIA

The number 142,857 is called a **cyclic number** and is one of the most remarkable integers. When multiplied by any number from 1 to 6 the result always consists of the same digits and, these digits always appear in the same cyclic order (with each number commencing at a different point).

$$
\begin{aligned}
1 \times 142{,}857 &= 142{,}857\\
2 \times 142{,}857 &= 285{,}714\\
3 \times 142{,}857 &= 428{,}571\\
4 \times 142{,}857 &= 571{,}428\\
5 \times 142{,}857 &= 714{,}285\\
6 \times 142{,}857 &= 857{,}142
\end{aligned}
$$

4.16 SUM OF FACTORIALS

Look at the following factorial relationships.

$$
\begin{aligned}
1 &= 1!\\
2 &= 2!\\
145 &= 1! + 4! + 5!\\
40{,}585 &= 4! + 0! + 5! + 8! + 5!
\end{aligned}
$$

A college student in North Carolina made an exhaustive search with a microcomputer and found the above factorial relationships.

REVIEW EXERCISES

1. Write a program that will find three integers, each less than 100, that give the same result when multiplied or added.

$$X \times Y \times Z = X + Y + Z$$

2. Write a program that will find all positive integers X less than 2000 that will make X(X + 180) a square. Your program should produce the answer: 12, 16, 60, 144, 320, 588 and 1936.

3. Modify the SUM AND PRODUCT program to find the numbers that have the property I + J + K + L = I × J × K × L. Your program should produce just one set of digits: 1, 1, 2, and 4, whose permutations generate twelve solutions.

4. Some numbers can be written as the sum of three squares. For instance,

$$43 = 5^2 + 3^2 + 3^2$$

Write a program that will find numbers that can be written as the sum of three squares.

5. Write a program to find three positive integers whose sum is 25 and whose product is 540.

6. The number $(ABBBB)^2 - 1$ has 10 digits, all different. Write a program that will produce at least two numbers. To check your program, use the following:

$$
\begin{aligned}
85555^2 - 1 &= 7319658024\\
97777^2 - 1 &= 1560341728
\end{aligned}
$$

7. Write a program to find a number that leaves the remainder 16 when divided by 39 and the remainder 27 when divided by 56.

I think it's another set of amicable numbers!

8. Addition seems to be the reverse of multiplication in the following:

$$9 + 9 = 18 = 9 \times 9 = 81$$
$$24 + 3 = 27 = 24 \times 3 = 72$$
$$47 + 2 = 49 = 47 \times 2 = 94$$
$$497 + 2 = 499 = 497 \times 2 = 994$$

Write a program to find other numbers with this property.

9. Write a program that will find integers that satisfy the relationship

$$X^2 + Y^2 = Z^2 + 1$$

10. Write a program that will find the smallest integers satisfying the equation

$$X^2 - 1620Y^2 = 1$$

Your program should have produced the answer: X = 161 and Y = 4.

11. The square of the sum of the first N consecutive integers is equal to the sum of the cubes of these integers, that is,

$$(1 + 2 + 3 + \ldots + N)^2 = 1^3 + 2^3 + 3^3 + \ldots + N^3.$$

Write a program that will prove this relationship for values of N up to 20. You may want to use the following formula. Since $1 + 2 + 3 + \ldots + N$ is an arithmetic progression whose sum is $N(N + 1)/2$, it follows that

$$1^3 + 2^3 + 3^3 + \ldots + N^3 = (N(N + 1)/2)^2$$

NUMBER THEORY TRIVIA

A prime number such as 11 is called a **repunit** (a number made up of repeating "units" or "ones"). Another large repunit is

1,111,111,111,111,111,111

which has nineteen ones. Two other large repunits have 23 and 317 ones. It is not known whether prime numbers of this form exist to large values. Professor A.H. Beiler is the originator of the word "repunit."

12. Write a program to generate number arrangements that satisfy the relationship discussed in Section 4.11.

13. Write a program that will find the smallest integers satisfying the equation

$$X^2 - 1666Y^2 = 1$$

Your program should have produced the answer X = 2449 and Y = 60.

14. Write a program that will find an integer less than 1000 whose cube could be represented in at least 5 distinct ways as the sum of the cubes of 3 positive integers. That is, find $N^3 = A^3 + B^3 + C^3$ where (A, B, C) represent at least 5 distinct sets of values and N is less than 1000. One solution to this problem is N = 492.

A	B	C
24	204	480
48	85	491
72	384	396
113	264	463
144	360	414
176	204	472
207	297	438
226	332	414
246	328	410
281	322	399

15. Modify the Sum Of Three Cubes program so that it will compute all 9-digit numbers with this property. For example

$$166{,}500{,}333 = 166^3 + 500^3 + 333^3$$

16. The following numbers have an interesting relationship.

$$135 = 1^1 + 3^2 + 5^3$$
$$175 = 1^1 + 7^2 + 5^3$$
$$518 = 5^1 + 1^2 + 8^3$$

Write a program that will produce all three digit numbers that have this relationship.

17. The prime number 47 is the sum of four squares (even in two ways).

$$47 = 6^2 + 3^2 + 1^2 + 1^2$$
$$= 5^2 + 3^2 + 3^2 + 2^2$$

Write a program that will find numbers that can be written as the sum of four squares.

5

FACTORING

PREVIEW

In Chapter 2 we discussed prime numbers with little reference to factors. The two are closely related since most methods of finding primes also infer factorization. For example, the number 12 is not prime because the factors of 12 are

1, 2, 3, 4, 6, and 12.

Factors, then, are those integers that can divide into a number exactly.

After completing this chapter, you should be able to:

1. Identify the prime factors, the greatest common factor, and the least common multiple of numbers.
2. See how a microcomputer can be used to determine the prime factors, the greatest common factor and the least common multiple of numbers.
3. Identify a procedure for finding the greatest common factor and the least common multiple of numbers.
4. Write programs to calculate the factors of a number.

5.1 PRIME FACTORIZATION

In Chapter 2 we saw that every integer greater than 1 is either a prime number or a composite number. In this section we shall find that every integer greater than 1 can be expressed in terms of its prime factors essentially in only one way.

The operation of factoring is the reverse of the operation of multiplying. Multiplying 3 by 6 yields

$$3 \times 6 = 18$$

and the answer is unique (only one answer is possible). The reverse process is called factoring. Suppose we are given the number 18 and are asked for numbers that can be multiplied together to give 18. We can then list the different factorization of 18:

$$18 = 2 \times 9$$
$$18 = 3 \times 6$$
$$18 = 2 \times 3^2$$

Notice that the last factorization contains only prime factors; thus it is called the prime factorization of 18.

It should be clear that, if a number is composite, it can be factored into two counting numbers greater than 1. These two numbers themselves will be prime or composite. If they are prime, then we have a prime factorization. If one or more is composite, we repeat the process. This continues until we have represented the original number as a product of primes. It is also true that this representation is unique.

Example 1: Find the prime factorization of 385. One of the easiest ways of finding the prime factors of a number is to try division by each of the prime numbers in order: 2, 3, 5, 7, 11, 13, 17 and 19. If none of these primes divides 385, then 385 must be a prime number. We see by inspection that 385 is not divisible by 2 or 3. It is divisible by 5, so

$$385 = 5 \times 77$$

Since 77 is a composite number that is divisible by 7 and 11, we write

$$385 = 5 \times 7 \times 11$$

Example 2: Find the prime factorization of 1400

$$1400 = 2 \times 2 \times 2 \times 5 \times 5 \times 7 = 2^3 \times 5^2 \times 7$$

Example 3: Find the prime factorization of 4680.

$$4680 = 2 \times 2 \times 2 \times 3 \times 3 \times 5 \times 13 = 2^3 \times 3^2 \times 5 \times 13$$

Example 4: Find the prime factorization of 493.

$$493 = 17 \times 29$$

NUMBER THEORY TRIVIA

The only standing ovation ever received at a meeting of the American Mathematical Association was given to a mathematician named Frank Cole in 1903. It was commonly believed that $2^{67} - 1$ was a prime. Cole first multiplied out 2^{67} and then subtracted 1. Moving to another board, he wrote:

$$\begin{array}{r} 761,838,257,287 \\ \times\ 193,707,721 \\ \hline \end{array}$$

He multiplied it out and came up with the same result as on the first blackboard. He had factored a number thought to be a prime!.

The following program will find the prime factors of any integer. The integer is input to the program as data. The program terminates execution whenever a zero is typed as input.

```
program PrimeFactors;
var
  B,A,N,I       :Integer;
begin
  writeln('PRIME  FACTORS  OF  ANY  INTEGER');
  writeln;
  write('NUMBER  TO  BE  FACTORED  IS ');
  readln(A);
  N := A;
  (**** Compute prime factors ****)
  B := 0;
  for I := 2 to (N div 2 + 1) do
    begin
      if (N/I) <= (N div I) then
        begin
          B := B + 1;
          if B <= 1 then writeln('PRIME FACTORS OF ',N, ' ARE');
          write(I, '    ');
          N :=  N div I;
          if N = 1 then HALT;
          I := I - 1;
        end;
    end;
  if N = A then
    writeln(N, ' IS A PRIME NUMBER');
end.
```

```
COMPILE/RUN

PRIME  FACTORS  OF  ANY  INTEGER


NUMBER  TO  BE  FACTORED  IS  638
PRIME  FACTORS  OF  638  ARE
2  11  29

NUMBER  TO  BE  FACTORED  IS  150
PRIME  FACTORS  OF  150  ARE
2  3  5  5

NUMBER  TO  BE  FACTORED  IS  71
71  IS  A  PRIME  NUMBER

NUMBER  TO  BE  FACTORED  IS  86791
PRIME  FACTORS  OF  86791  ARE
229  379
```

NUMBER THEORY TRIVIA

Factoring large numbers is difficult. It takes a day or so of a supercomputer's time to factor an 80-digit number that happens to have no small factor. Consider a number like $2^{193} - 1$. The smallest prime factor is 13,821,503. The second prime divisor lies somewhere in more distant parts. In fact, if a computer could perform a billion division instructions per second, it would require more than 35,000 years of computer time to find the second largest factor of this number. Mathematicians Carl Pomerance and Samuel Wagstaff found a special factoring method and determined that

$2^{193} - 1 =$
13,821,503 ×
61,654,440,233,248,340,616,559 ×
14,732,265,321,145,317,331,353,
282,383

with each factor a prime.

In the sample program runs, four integers are factored, including one prime number and one integer with two fairly large prime factors. Notice in the second example that 5 appears twice, indicating that it is a multiple factor. In the third example a prime number is tested (71).

In 1643, Pierre de Fermat (1601-1665), a French mathematical genius, illustrated a method of factoring numbers that did not require divisions. His method was based on the following procedure:

Assume	$N = A * B$ where $A \leq B$.
Assume	N, A, and B are all odd integers.
Let	$X = (A + B)/2$ and,
	$Y = (B - A)/2$.
Then	$N = X^2 - Y^2$.

The method involves searching for values of A and B that satisfy these equations where $0 < B < A$ and $A < B \leq N$.

The following program computes the largest factor of a given integer.

```
program LargestFactor;
var
  N      :Integer;(* Number *)
  D      :Integer;(* Factor of number *)
begin
  write('ENTER AN INTEGER LARGER THAN ZERO ');
  readln(N);
  (**** Find the largest factor of integer N ****)
  for D := 2 to TRUNC(SQRT(N)) do
    if N/D = N div D then
      begin
        writeln((N div D):6, ' IS THE LARGEST FACTOR OF ',N);
        HALT;
      end;
  (**** Integer could not be factored ****)
  writeln(N,' IS A PRIME NUMBER ');
end.
```

```
COMPILE/RUN

ENTER AN INTEGER LARGER THAN ZERO 373
373 IS A PRIME NUMBER

ENTER AN INTEGER LARGER THAN ZERO 1495
299 IS THE LARGEST FACTOR OF 1495

ENTER AN INTEGER LARGER THAN ZERO 5799
1933 IS THE LARGEST FACTOR OF 5799
```

In October of 1988, a team of mathematicians and computer scientists managed, after 26 days of computation, to factor a 100-digit number into two large prime factors, 41-digits and 60-digits long. The process involved an indirect factoring strategy and utilized a world-wide network of computers, each of which worked on small portions of the project. The various independent results were relayed by electronic mail to the D.E.C. Systems Research Center in California for the final analysis.

To find all pairs of factors of an integer, use the following program.

```
program Factors;
var
  X,A        :Integer;
begin
  writeln('PAIRS OF FACTORS');
  writeln;
  writeln('ENTER THE INTEGER: ');
  readln(X);
  writeln;
  writeln('THE PAIRS OF FACTORS OF ',X, ' ARE: ');
  for A := 1 to TRUNC(SQRT(ABS(X))) do
    if X div A = X / A then writeln(A, '          ', X div A);
end.
```

```
COMPILE/RUN
PAIRS OF FACTORS

ENTER THE INTEGER: 2656

THE PAIRS OF FACTORS OF 2656 ARE:
 1          2656
 2          1328
 4          664
 8          332
 16         166
 32         83
```

NUMBER THEORY TRIVIA

In 1988, a group of mathematicians at the University of Georgia factored a 95-digit number. The number occurs as the remainder when $2^{332} + 1$ is divided by 17 and 11,953. While this effort set no record in terms of time or digits, it was significant in that all but the final computation was done using about 100 microcomputer systems with the data transfer accomplished by hand-carrying of floppy disks from microcomputer to microcomputer.

5.2 GREATEST COMMON DIVISOR

The problem of factoring is closely related to finding the greatest common divisor (GCD) of two numbers.

If A and B are two integers, any number that divides both A and B is called a common divisor of A and B. For example, the numbers that divide 12 are 1, 2, 3, 4, 6, and 12. The numbers that divide 18 are 1, 2, 3, 6, 9, 18. Then the common divisors of 12 and 18 are 1, 2, 3, 6. The largest of the divisors is called the greatest common divisor (GCD) of A and B. In the previous example, 6 is the greatest common divisor (GCD) of 12 and 18.

NUMBER THEORY TRIVIA

Strobogrammatic numbers are numbers that read backwards after having been rotated 180°; e.g., 69, 96, 1001.

We may use the prime factorization of two numbers to find their greatest common divisor. Express each number by its prime factorization, consider the prime numbers that are factors of both of the given number, and take the product of those prime numbers with each raised to the highest power that is a factor of both of the given numbers. For $12 = 2^2 \times 3$ and $18 = 2 \times 3^2$ we have GCD = 2×3; that is, 6.

Example 1. Find the greatest common divisor of 3850 and 5280.

$$3850 = 2 \times 5^2 \times 7 \times 11;$$
$$5280 = 2^5 \times 3 \times 5 \times 11.$$

The GCD of 3850 and 5280 is $2 \times 5 \times 11$; that is, 110.

Even though the GCD of two numbers can be found quite easily by using the previous method, there exists an ancient method that is more suited for computer computation. A procedure known as Euclid's algorithm is one of the basic methods of elementary number theory. Suppose that A and B are the two numbers. We divide B into A, using integer division, finding a quotient Q and a remainder R. This means that

$$A = Q \times B + R$$

Then we divide R into B and keep iterating until the remainder is 0. The last nonzero divisor becomes the greatest common divisor.

Example 2. Let us find the GCD of 1976 and 1032.

$$1976 = 1032 \times 1 + 944$$
$$1032 = 944 \times 1 + 88$$
$$944 = 88 \times 10 + 56$$
$$88 = 56 \times 1 + 32$$
$$56 = 32 \times 1 + 24$$
$$32 = 24 \times 1 + 8$$
$$24 = 8 \times 3 + 0$$

The GCD of 1976 and 1032 is 8.

Example 3. Find the GCD of 76084 and 63020.

$$76084 = 63020 \times 1 + 13064$$
$$63020 = 13064 \times 4 + 10764$$
$$13064 = 10764 \times 1 + 2300$$
$$10764 = 2300 \times 4 + 1564$$
$$2300 = 1564 \times 1 + 736$$
$$1564 = 736 \times 2 + 92$$
$$736 = 92 \times 8 + 0$$

The GCD of 76084 and 63020 is 92.

The following program uses Euclid's algorithm to compute the GCD of a given pair of numbers.

```
program GreatestCommonDivisor;
var
  Number1, Number2, Remainder      :Integer;
begin
  write('ENTER FIRST NUMBER: ');
  readln(Number1);
  write('ENTER SECOND NUMBER: ');
  readln(Number2);
  write('THE GCD OF ', Number1, ' AND ', Number2);
  repeat
    Remainder := Number1 mod Number2;
    if Remainder <> 0 then
      begin
        Number1 := Number2;
        Number2 := Remainder;
      end;
  until Remainder = 0;
  (**** Print the GCD of numbers ****)
  write(' IS ', Number2);
end.
```

COMPILE/RUN

```
ENTER FIRST NUMBER: 60
ENTER SECOND NUMBER: 5280
THE GCD OF 60 AND 5280 IS 60
```

NUMBER THEORY TRIVIA

The following prime numbers have the three-digit endings 111, 333, 777, and 999.

2111	2333
79111	79333
136111	136333
2777	2999
79777	79999
136777	136999

Can you write a program to produce additional primes with these three-digit endings?

NUMBER THEORY TRIVIA

A group of researchers at Purdue University and the University of Georgia are building their own computer which is designed specifically to do factoring, and only factoring. The machine is called EPOC (for Extended Precision Operand Computer), however, it has a nickname, "The Georgia Cracker." The expectation is that most numbers up to 80-digits in length will be within reach of EPOC using computational times of less than one day.

5.3 LEAST COMMON MULTIPLE

The set of multiples of 12 is 12, 24, 36, 48, 60, 72, 84, 96, 108, . . .; the set of multiples of 18 is 18, 36, 54, 72, 90, 108, . . .; the set of common multiples of 12 and 18 is 36, 72, 108, . . . Notice that the least of these common multiples, 36, is a divisor of each of the common multiples. In general, the common multiple of two numbers which is a divisor of each of the common multiples is called the least common multiple (LCM) of the two numbers.

We may use the prime factorization of two numbers to find their least common multiple. Express each number by its prime factorization, consider the prime factors that are factors of either of the given numbers, and take the product of these prime numbers with each raised to the highest power that occurs in either of the prime factorizations. For $12 = 2^2 \times 3$ and $18 = 2 \times 3^2$, we have LCM $= 2^2 \times 3^2$; that is, 36.

Example 1. Find the least common multiple of 3850 and 5280.

$$3850 = 2 \times 5^2 \times 7 \times 11;$$
$$5280 = 2^5 \times 3 \times 5 \times 11.$$

The least common multiple of 3850 and 5280 is $2^5 \times 3 \times 5^2 \times 7 \times 11$; that is 184,800.

A common multiple of three numbers A, B, and C is a number divisible by all of them. Among these multiples there is a least common multiple. The LCM of 2, 3, and 6 is 6. The LCM of 5, 10, and 20 is 20. The LCM of 12, 18, and 24 is 72. A general technique for finding the LCM of three numbers follows:

1. Identify the numbers A, B, C.
2. Arbitrarily let X = A.
3. Does A divide X?
4. If yes, go to Step 6.
5. If no, increase X by 1 and return to Step 3.
6. Does B divide X?
7. If yes, go to Step 9.
8. If no, increase X by 1 and return to Step 3.
9. Does C divide X?
10. If yes, go to Step 12.
11. If no, increase X by 1 and return to Step 3.
12. Print the value of X.

Example 2. Find the least common multiple of 300, 144, and 108.

$$300 = 2^2 \times 3^1 \times 5^2$$
$$144 = 2^4 \times 3^2 \times 5^0$$
$$108 = 2^2 \times 3^3 \times 5^0$$

The least common multiple = $2^4 \times 3^3 \times 5^2 = 16 \times 27 \times 25 =$ 10,800. This means that the smallest number that 300, 144, and 108 all divide into is 10,800.

A program to compute the least common multiple of three numbers follows.

```
program LeastCommonMultiple;
var
  A,B,C,X      :Integer;
begin
  write('ENTER THREE INTEGERS: ');
  readln(A,B,C);
  X := A;
  while X < 32767 do
    begin
      if X div A = X / A then
        if X div B = X / B then
          if X div C = X / C then
            begin
              writeln('THE LCM OF ',A, ' ',B, ' ',C, ' IS ',X);
              HALT;
            end;
      X := X + 1;
    end;
  writeln('Program cannot compute the LCM of these integers');
end.
```

```
COMPILE/RUN

ENTER THREE INTEGERS: 5,10,20
THE LCM OF 5    10    20    IS    20
```

NUMBER THEORY TRIVIA

113797 is an interesting prime number. Write it backwards (797311) and it is another prime. Write it as two successive three-digit numbers (113 797 or 797 311) and both are three-digit reversible numbers. Change the 113 to 131 (131797) and it becomes another reversible six-digit prime, 797131. And both of the three-digit primes are palindromes.

5.4 FINDING THE GCD AND LCM OF SEVERAL NUMBERS

The greatest common divisor (GCD) will be less than or equal to the numbers involved. The least common multiple (LCM) will be greater than or equal to the numbers involved.

So far the greatest common divisor and the least common multiple have been explained only for two and three numbers, respectively, but there is no difficulty in extending these concepts. Let us

Several of the most secure cipher systems are based on the fact that large numbers are extremely difficult to factor, even using the most powerful computers for long periods of time. One cryptographic system based on the difficulty of factoring large numbers was invented in 1977 by Ronald L. Rivest, Adi Shamir and Leonard Adelman, all of the Massachusetts Institute of Technology. In the RSA System (which takes its initials from the names of its inventors), digits replace each letter in a text message, and the entire sequence of digits is treated as a single large number. A mathematical operation is then performed on this number, and to decipher the result requires either that the receiver possesses the key or breaks the code by factoring the large number.

consider first the case of three numbers A, B, and C. A common divisor is any number dividing them all. Among these common divisors there is a greatest common divisor. For example, let us determine the GCD of the three numbers 76,084, 63,020, and 196. In a previous example we have already found that the GCD of 76,084 and 63,020 is 92; consequently the GCD of 92 and 196 is 4. Therefore, the GCD of 76,084, 63,020, and 92 is 4.

To find the LCM of the three numbers 24, 18, and 52, you calculate the LCM of 24 and 18, which is 72, and then the LCM of 72 and 52, which results in 936. Therefore, the LCM of 24, 18, and 52 is 936.

All the properties associated with producing the GCD and LCM for three numbers can be used to produce the GCD and LCM for any set of numbers.

```
program GCDAndLCMOfFourNumbers;
var
  A                :array[1..4] of Integer;
  B                :array[1..4] of Integer;
  L,G,I,J,K,F,M    :Integer;
begin
  writeln('THIS PROGRAM COMPUTES THE LCM AND');
  writeln('GCD OF FOUR INTEGERS');
  write('TYPE THE FOUR INTEGERS: ');
  readln(A[1], A[2], A[3], A[4]);
  K := A[1];
  (**** Find largest number ****)
  for I := 1 to 4 do
    if K < A[I] then K := A[I];
  L := 1;
  G := 1;
  (**** Compute the LCM and GCD ****)
  for J := 2 to K do
    begin
      for I := 1 to 4 do
        B[I] := 1;
      I := 1;
      while I < = 4 do
        begin
          if A[I] / J = INT(A[I] / J) then
            begin
              A[I] := A[I] div J;
              B[I] := B[I] * J;
              I := I - 1;
            end;
          I := I + 1;
        end;
      F := B[1];
      M := B[1];
```

NUMBER THEORY TRIVIA

If a number is a multiple of 3, then the same number with the digits reversed is also a multiple of 3. For example, since 15 is a multiple of 3 we know that 51 is also. Multiples of 33 have this property too. For example, since 2244 is a multiple of 33, so must 4422.

```
        for I := 2 to 4 do
          begin
            if F < B[I] then F := B[I];
            if M > B[I] then M := B[I];
          end;
        L := L * F;
        G := G * M;
      end;
    writeln;
    writeln('THE LEAST COMMON MULTIPLE IS ',L);
    writeln('THE GREATEST COMMON DIVISOR IS ',G);
end.
```

```
COMPILE/RUN

THIS PROGRAM COMPUTES THE LCM AND
GCD OF FOUR INTEGERS
TYPE THE FOUR INTEGERS: 2 6 8 12
THE LEAST COMMON MULTIPLE IS 24
THE GREATEST COMMON DIVISOR IS 2
```

REVIEW EXERCISES

1. Write the prime factorization for each of the following numbers.

a. 28	d. 215	g. 300
b. 76	e. 143	h. 820
c. 130	f. 51	i. 97

2. Indicate which of the following numbers are prime. If the number is not prime, write its prime factorization.

a. 263	c. 445	e. 913
b. 323	d. 877	f. 1279

3. Find the prime factorization of 101,101.
4. Using Euclid's algorithm, determine the GCD of 15 and 28.
5. Using Euclid's algorithm, determine the GCD of 68 and 76.
6. Find the GCD of the following sets of numbers.

a. 60 and 72	c. 95 and 1425
b. 9, 12 and 14	d. 12, 54 and 171

7. Find the GCD of each of the first four pairs of amicable numbers. Check the results against those obtained from the prime factorizations.
8. Find the LCM of the following sets of numbers.

a. 60 and 72	c. 95 and 1425
b. 9, 12 and 14	d. 12, 54 and 171

9. Find the LCM for each of the first four pairs of amicable numbers.
10. Find the GCD and LCM of 75, 120 and 975.

NUMBER THEORY TRIVIA

Permutable primes are the prime numbers with at least two distinct digits which remain prime no matter how we rearrange (permute) their digits.

13 31
17 71
37 73
79 97
113 131 311

Can you write a program to find additional permutable primes?

NUMBER THEORY TRIVIA

A **distinct-digit sum** is an addition using nine digits (1 through 9) or ten digits (0 through 9) without repetition:

286 + 173 = 459

587 + 437 = 1026

11. a. Find the GCD of 6 and 15
    b. Find the LCM of 6 and 15
    c. Find the product of your answers to parts a and b.
    d. Compare your answer to part c with the product of 6 and 15.
12. The number 10 has four factors: 1, 2, 5, and 10. They all divide evenly into 10. The number 48 has ten factors: 1, 2, 3, 4, 6, 8, 12, 16, 24, and 48. They all divide evenly into 48. Write a program to compute the smallest natural number that has exactly 32 factors.
13. Write a program to find the sum of all numbers between 100 and 1000 that are divisible by 14.
14. Write a program to determine what value of X would make X434X0 divisible by 36.
15. Write a program to compute and print the LCM and GCD of several sets of six numbers.

6

FIBONACCI NUMBERS

PREVIEW

The great mathematician of the Middle Ages was Leonardo of Pisa, called Fibonacci. The construction of the famous Leaning Tower of that city was begun during his lifetime but was not completed for nearly two centuries. Fibonacci studied in North Africa, where he learned mathematical works available only in Arabic. In 1202, he wrote a book on arithmetic and algebra titled the **Liber Abaci**. In this book, he proposed a problem about rabbits whose solution was based on the number sequence

1 1 2 3 5 8 13 . . .

Fibonacci, or Leonardo of Pisa (1170-1250), famous 12th and early 13th century Italian mathematician. He was generally regarded as the greatest and most productive mathematician of the Middle Ages.

This sequence is called the Fibonacci sequence. It appears in an amazingly wide variety of creations, both natural and peoplemade.

After completing this chapter, you should be able to:

1. Identify number patterns and sequences, the Fibonacci number sequence, the golden ratio and the Fibonacci sorting technique.
2. Identify a formula for producing the Fibonacci number sequence.
3. See how a microcomputer can be used to generate Fibonacci numbers.
4. Identify several uses for Fibonacci numbers.
5. Write a program to generate Fibonacci numbers.

n	F_n
1	1
2	1
3	2
4	3
5	5
6	8
7	13
8	21
9	34
10	55
11	89
12	144
13	233
14	377
15	610
16	987
17	1,597
18	2,584
19	4,181
20	6,765
21	10,946
22	17,711
23	28,657
24	46,368
25	75,025
26	121,393
27	196,418
28	317,811
29	514,229
30	832,040
31	1,346,269
32	2,178,309
33	3,524,578
34	5,702,887
35	9,227,465
36	14,930,352
37	24,157,817
38	39,088,169
39	63,245,986
40	102,334,155

The first forty Fibonacci numbers.

6.1 MATHEMATICAL PATTERNS

A number sequence or progression is a collection of numbers arranged in order so that there is a first term, a second term, a third term, and so on.

Example 1. What is the pattern for the sequence 1, 4, 7, 10, 13, . . .? Fill in the next three terms.

There is a difference of 3 between each two terms. That is, 3 is added to each term in order to find the next term. The next three terms are 16, 19, and 22.

If the difference between each two terms of a sequence is the same, it is called the common difference. The common difference in Example 1 is 3. If a sequence has a common difference, it is called an arithmetic sequence.

Example 2. Which of the following sequences are arithmetic?

a. 6, 9, 12, 15, . . . b. 66, 72, 78, 84, . . .
c. 4, 11, 18, 24, . . . d. 113, 322, 531, 740, . . .

a. The common difference is 3; the next term is found by adding 3: 15 + 3 = 18.

b. The common difference is 6; the next term is 84 + 6 = 90.

c. There is no common difference, so this sequence is not arithmetic.

d. The common difference is 209; the next term is 740 + 209 = 949.

Example 3. What is the pattern for the sequence 2, 4, 8, 16, 32, . . .? Fill in the next three terms. This is not an arithmetic sequence because there is no common difference. However, if each term is divided by the preceding term, the results are the same. This number found by division is called a common ratio. If a sequence has a common ratio, then the sequence is called a geometric sequence. Additional terms of a geometric sequence can be found by multiplying successive terms by the common ratio. For this example the common ratio is 2, so the next three terms are 64, 128 and 256.

Example 4. Give the next term for the sequence A, C, F, J, O, . . .

Not all patterns need be patterns of numbers. It is, of course, not arithmetic or geometric, but notice that after A there is one letter (B) omitted; then C; then two letters omitted; then F; then three letters omitted; J; four letters omitted; O; so the next gap should have five letters omitted; thus, the next letter in the pattern is U.

This must be Fibonacci's apartment!

Example 5. What is the next term for the sequence 15, 30, 45, 75, . . .?

The pattern is to add successive terms. This is a fairly common way of formulating a sequence.

15 + 30 = 45   30 + 45 = 75   45 + 75 = 120 (the next term)

The pattern of adding successive terms, as illustrated in this example, suggests a particular famous sequence that has a special name. This sequence is discussed in the next section.

NUMBER THEORY TRIVIA

Leonardo Fibonacci studied calculation with an Arab master while his father was serving as a consul in North Africa. His first and best known work is *Liber Abaci* (Book of the Abacus), published in 1202, and revised in 1228. In it he made the Arabic numeral system generally available in Europe. His *Practica Geometriae* (Practice of Geometry) appeared in 1220, and *Liber Quadratorum* (Book of Square Numbers) in 1225.

6.2 THE FIBONACCI SEQUENCE

A man bought a pair of rabbits and bred them. The pair produced one pair of young after one month, and a second pair after the second month. Then they stopped breeding. Each new pair also produced two more pairs in the same way, and then stopped breeding. How many new pairs of rabbits did he get each month?

To answer this question, write down in a line the number of pairs in each generation. First write the number 1 for the single pair he started with. Next write the number 1 for the pair they produced after a month (see Figure 6-1).

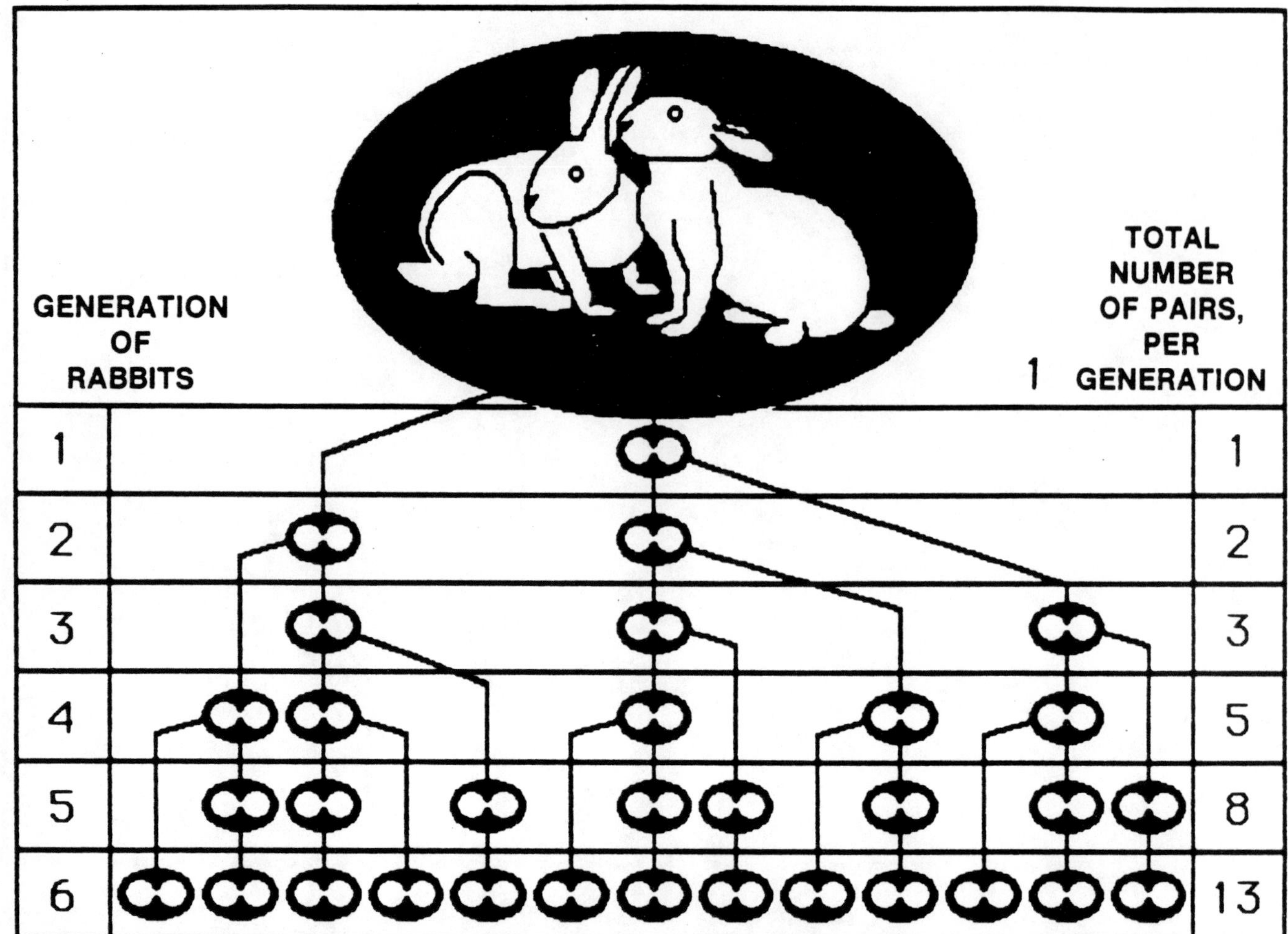

Figure 6-1
Rabbits and generations.

The next month both pairs had young, so the next number is 2. We now have three numbers in a line: 1, 1, 2, each number representing a new generation. Now the first generation stopped producing. The second generation (1 pair) produced 1 pair, and the third generation (2 pairs) produced 2 pairs, so the next number we write is 1 + 2, or 3. Now the second generation stopped producing. The third generation (2 pairs) produced 2 pairs, and the fourth generation (3 pairs) produced 3 pairs, so the next number we write is 2 + 3 or 5.

Each month, only the last two generations produced, so we can get the next number by adding the last two numbers in the line. The numbers we get this way are called Fibonacci numbers. The first sixteen are 1, 1, 2, 3, 5, 8, 13, 21, 34, 55, 89, 144, 233, 377, 610, and 987.

The Fibonacci numbers are such that, after the first two, every number in the sequence equals the sum of the two previous numbers:

$$F_n = F_{n-1} + F_{n-2}$$

NUMBER THEORY TRIVIA

Leonardo Fibonacci is a man of many names. He has also been called Leonardo Pisano, Leonardo the Pisan, Leonardo of Pisa, son of Fibonacci, son of Bonacci, son of Bonacius. His name, I. pisano, appears in one of his early manuscripts. He also used the nickname Bigollo.

Now I know why Fibonacci bought a pair of rabbits!

NUMBER THEORY TRIVIA

Every fifth number in the Fibonacci number sequence is divisible by 5. Can you write a program to illustrate this statement for the first 125 Fibonacci numbers?

Suppose a tree grows according to the following, not unrealistic, formula. Each old branch (including the trunk) puts out one new branch per year; each new branch grows through the next year without branching, after which it qualifies as an old branch. The growth is represented schematically in Figure 6-2. The number of branches after n years is F_n.

The Fibonacci numbers have very interesting properties, and keep popping up in many places in nature and art.

Here is one of the curious properties of these numbers. Pick any three numbers that follow each other in the line. Square the middle number and multiply the first number by the third number. The

Figure 6-2
The Fibonacci tree.

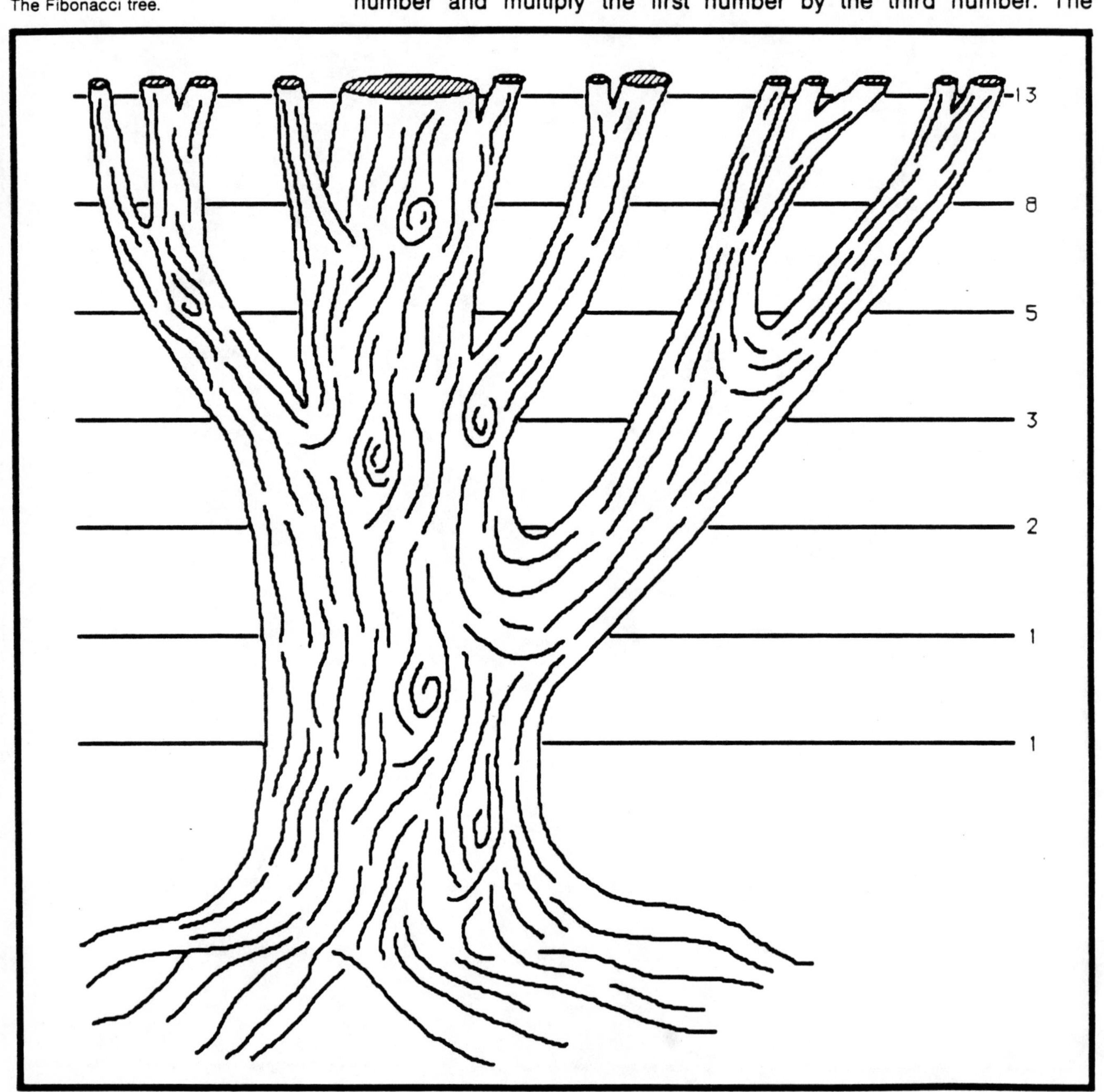

The Fibonacci sequence occurs in many forms of plant and animal life, such as in arrangements of parts into opposing spirals which correspond to consecutive Fibonacci numbers; for example, one spiral may contain five parts, and the opposing spiral eight parts. Such arrangements are found in pine cones (5:8), and daisy disks (21:34).

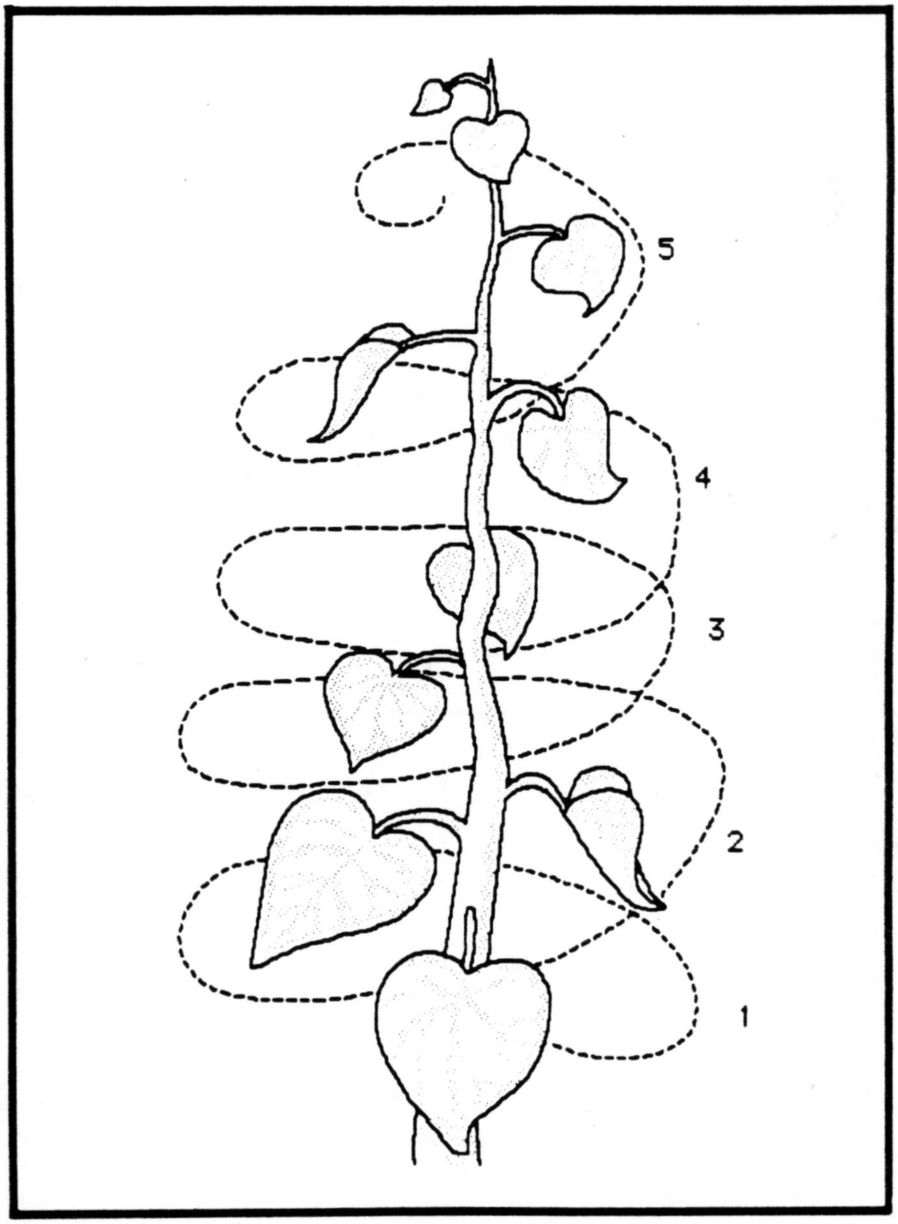

results will always differ by 1. For example, if we take the numbers 3, 5, 8, we get $5^2 = 5 \times 5 = 25$, while $3 \times 8 = 24$.

If we divide each number by its right hand neighbor, we get a series of fractions:

1/1 1/2 2/3 3/5 5/8 8/13 13/21 21/34 34/55

These fractions are related to the growth of plants. When new leaves grow from the stem of a plant, they spiral around the stem. The spiral turns as it climbs (Figure 6-3). The amount of turning from one leaf to the next is a fraction of a complete rotation around the stem. *This fraction is always one of the Fibonacci fractions.* Nature spaces the leaves in this way so that the higher leaves do not shade the lower leaves too much.

Figure 6-3
In the example above there are five complete turns from leaf 1 to leaf 9. The Fibonacci ratio for this plant is 5/8.

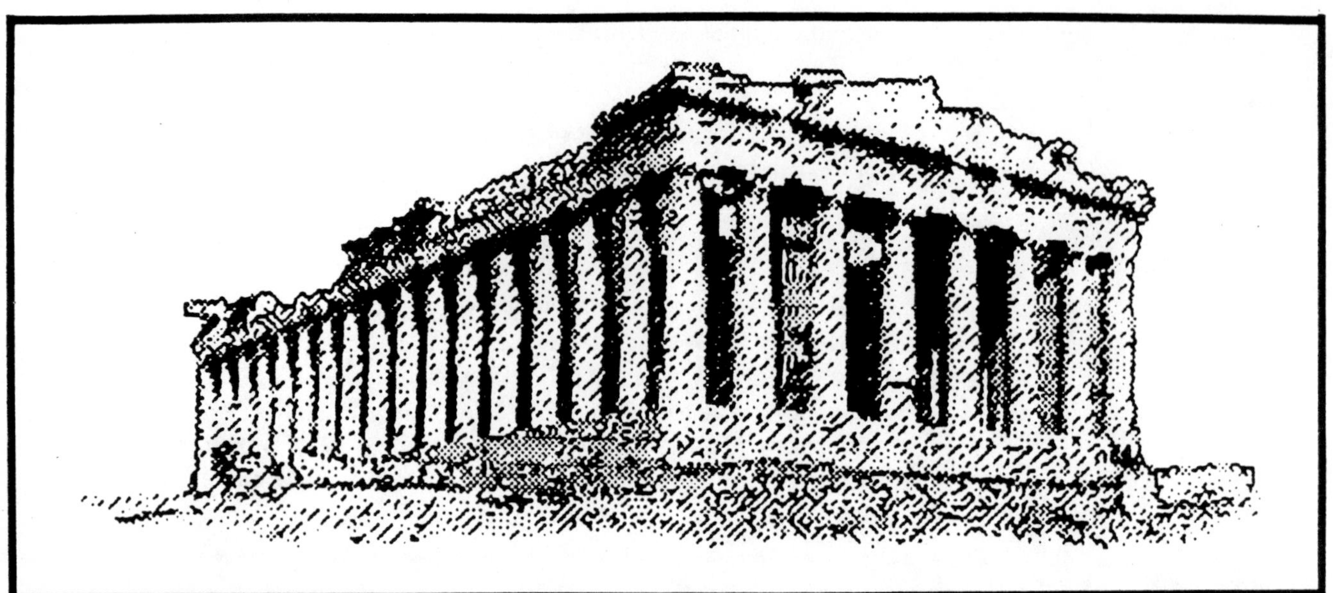

The ancient temple, the Parthenon, on the Acropolis in Athens, Greece, was built in the golden rectangle's proportions, although in the 5th century B.C. the builders probably did not know about it. The temple was built to honor Pallas Athene, the fabled beauty, who was the Goddess of Wisdom.

The same fractions come up in art. For example, not all rectangles are equally pleasing to the eye. Some look too long and narrow. A square looks too stubby and fat. There is a shape between these extremes that looks the best. In this best-looking of rectangles the ratio of the width to the length is about the same as the ratio of the length to the sum of the width and length. It is called the *golden section.*

There is a formula that gives directions for calculating the golden section. The directions are: Subtract 1 from the square root of 5, and divide by 2. The square root of 5 is approximately 2.24. Subtracting 1, and dividing by 2, we get .62 as an approximate value of the golden section. Fibonacci fractions are close to the golden section. In fact, the further out they are in the series, the closer they get to it. The fraction 5/8 is closer to the golden section than 3/5. The fraction 8/13 is closer than 5/8, and so on.

A program that computes and prints 23 Fibonacci numbers follows:

```
program FibonacciNumbers;
var
  FibNu     :array[1..23] of Integer;
  N         :Integer;
begin
  writeln('FIBONACCI NUMBERS');
  writeln;
  (**** Establish first two Fibonacci numbers ****)
  FibNu[1] := 1;
  FibNu[2] := 1;
  (**** Generate the next 21 Fibonacci numbers ****)
  for N := 1 to 21 do
    FibNu[N + 2] := FibNu[N + 1] + FibNu[N];
  (**** Print 23 Fibonacci numbers ****)
  for N := 1 to 23 do
```

From 1927 to 1934, some of the wildest years in Wall Street history, an accountant named Ralph Elliott charted the patterns of stock averages and commodities prices. To his surprise, the patterns were the same whether he charted trends of a few hour's duration or ones spanning years. He eventually identified nine trends and formulated the Elliott Wave Principle. The wave patterns of the price movements were identical to the mathematical progression found in the spiral pattern on pineapples - the Fibonacci series. One feature of the series is that after the first four or so numbers, the ratio of any number to the next one (21 to 34, say) in the sequence approaches .615 to 1, or, more simply, 3 to 5. This is the Fibonacci ratio. Elliott's waves are based on this ratio. That is, the prices he tracked followed a rhythm of five waves up and three waves down.

```
    writeln(FibNu[N]);
end.
```

COMPILE/RUN

```
FIBONACCI NUMBERS
1
1
2
3
5
8
13
21
34
55
89
144
233
377
610
987
1597
2584
4181
6765
10946
17711
28657
```

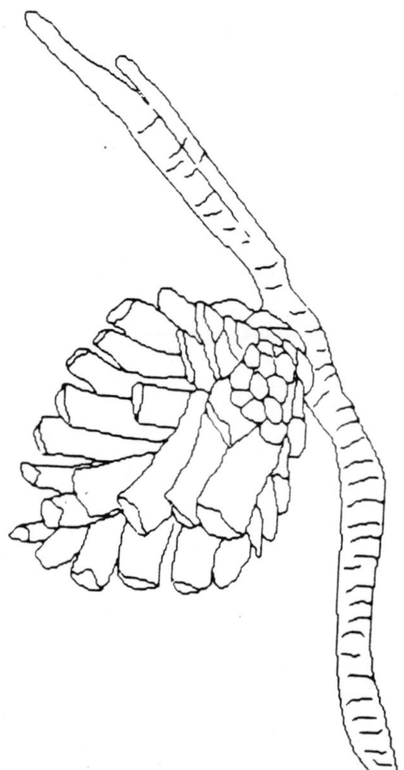

The Fibonacci numbers 5 and 8 occur in pine-cone scales.

NUMBER THEORY TRIVIA

Neighboring numbers in the **Fibonacci number sequence** are prime to each other (have no common factor).

6.3 FIBONACCI NUMBERS AND PRIMES

The program in the last section produced the first thirty Fibonacci numbers. As you have observed, all of these Fibonacci numbers are positive integer quantities, and some are prime numbers.

Let us now consider the problem of generating Fibonacci numbers and identifying those that are primes. An outline of the program procedure follows.

1. Set F1 and F2 to 1 (F1 is the first Fibonacci number and F2 is the second Fibonacci number).
2. Print F1 and F2, identifying each as a prime number.
3. Perform the following calculations for I = 3, 4, . . ., 25.
   a. Calculate a value for F using the formula F = F1 + F2.
   b. Test to see if F is a prime number.
   c. If F is a prime, identify it as such.
   d. Update F1 and F2 in preparation for calculating a new Fibonacci number (assign the current value of F1 to F2, then assign the current value of F to F1).

A program corresponding to the previous procedure follows.

```
program FibonacciAndPrimeNumbers;
var
  N            :Integer;(* Number of Fibonacci numbers to compute *)
  F,F1,F2      :Integer;(* Fibonacci numbers *)
  I            :Integer;(* Looping counter *)
  J            :Integer;(* Counter *)
  Flag         :Integer;(* Prime number indicator *)
begin
  write('ENTER NUMBER OF FIBONACCI NUMBERS ');
  readln(N);
  writeln;
  writeln('FIBONACCI AND PRIME NUMBERS');
  writeln('------------------------------------');
  (**** Set and print the first two Fibonacci numbers ****)
  F1 := 1;
  F2 := 1;
  writeln(F1,' (PRIME NUMBER)');
  writeln(F2,' (PRIME NUMBER)');
  (**** Compute Fibonacci numbers ****)
  for I :=  3 to N do
    begin
      Flag := 0;
      F := F1 + F2;
      for J := 2 to F - 1 do
        if F/J = F div J then Flag := 1;
      if Flag = 1
        then writeln(F)
        else writeln(F,' (PRIME NUMBER)');
      F2 := F1;
      F1 := F;
    end;
end.
```

NUMBER THEORY TRIVIA

The sum of any ten numbers in the **Fibonacci number sequence** is divisible by 11.

```
COMPILE/RUN

ENTER NUMBER OF FIBONACCI NUMBERS 23

FIBONACCI AND PRIME NUMBERS
----------------------------------------

1  (PRIME NUMBER)
1  (PRIME NUMBER)
2  (PRIME NUMBER)
3  (PRIME NUMBER)
5  (PRIME NUMBER)
8
13  (PRIME NUMBER)
21
34
55
89  (PRIME NUMBER)
144
233  (PRIME NUMBER)
377
610
987
1597  (PRIME NUMBER)
2584
4181
6765
10946
17711
28657  (PRIME NUMBER)
```

Notice that the program contains a nest of loops. The purpose of the inner loop (statements 280-320) is to determine whether or not each Fibonacci number is prime; the outer loop (statements 260-380) causes the desired sequence of Fibonacci numbers to be computed.

The program generates the first twenty-three Fibonacci numbers, nine of which are unique prime numbers.

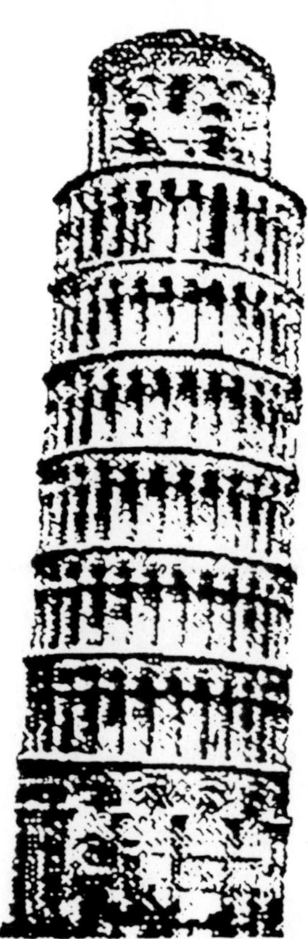

In the Italian city of Pisa there is a famous 185-foot tower of white marble that leans seventeen feet from the perpendicular. There is also another less spectacular monument in Pisa — the statue of Leonardo Fibonacci. He made major contributions to the development of modern mathematics.

6.4 THE GOLDEN RATIO

In a previous section, the Fibonacci sequence

$$1, 1, 2, 3, 5, 8, 13, 21, 34, 55, 89, 144, \ldots$$

was introduced. Suppose we consider the ratios of the successive terms of this sequence:

$$1/1 = 1.000$$
$$2/1 = 2.000$$
$$3/2 = 1.500$$
$$5/3 = 1.667$$
$$8/5 = 1.600$$

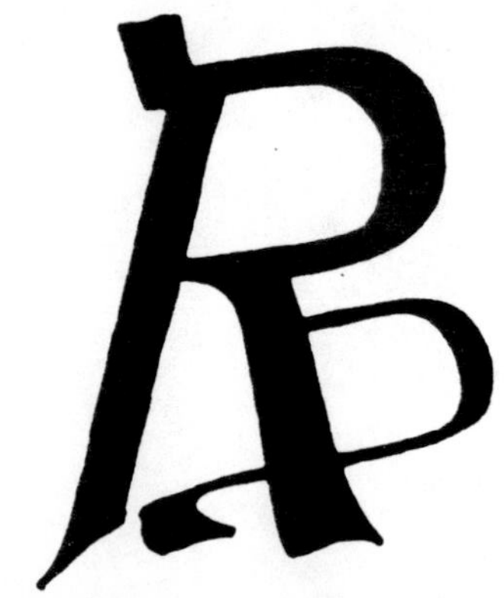

It is believed that Leonardo Fibonacci was the first person to use this sign in 1220. Coming from the Latin *radix* (meaning root), it is much more complex than today's radical sign ($\sqrt{\ }$), which is probably a distortion of the letter *r* from sixteenth century Germany.

13/8 = 1.625
21/13 = 1.615
34/21 = 1.619
55/34 = 1.618
89/55 = 1.618

If you continue finding these ratios, you will notice that the sequence oscillates about a number approximately equal to 1.618.

Suppose we repeat this same procedure with another sequence, this time choosing *any* two nonzero numbers, say 4 and 7. Construct a sequence by adding terms as was done with the Fibonacci sequence:

4, 7, 11, 18, 29, 47, 76, 123, 199, 322, . . .

Next, form the ratios of the successive terms:

7/4 = 1.750
11/7 = 1.571
18/11 = 1.636
29/18 = 1.611
47/29 = 1.621
76/47 = 1.617
123/76 = 1.618
199/123 = 1.618

These ratios are oscillating about the same number! There would seem to be something interesting about this number, which is called the **golden ratio** (1.6180339885).

The value of the golden ratio can be calculated in numerous ways. One method is to generate the Fibonacci number sequence and consider the ratio of each set of consecutive terms as it does so.

The following program computes the golden ratio for a specified number of terms in the Fibonacci sequence.

```
program GoldenRatio;
var
  A       :Integer;(* First Fibonacci number *)
  B       :Integer;(* Second Fibonacci number *)
  S       :Integer;(* Next Fibonacci number *)
  C       :Integer;(*Counter *)
  T       :Integer;(* Number of terms *)
begin
  C := 1;
  write('ENTER NUMBER OF TERMS ');
  readln(T);
  writeln('TERM A          TERM B            RATIO A/B');
  A := 1;
  B := 1;
```

NUMBER THEORY TRIVIA

The Golden Ratio used in art by Leonardo da Vinci et al, is derived from a special number sequence. Who discovered it?

a. Pascal
b. Euler
c. Fibonacci
d. Napier
e. Fermat

The Fibonacci Association was founded by mathematicians in 1960. The purpose of this organization is to exchange ideas and stimulate research on Fibonacci numbers and related topics. The organization publishes the *Fibonacci Quarterly* journal four times a year.

```
  repeat
    (**** Print Fibonacci numbers and ratio ****)
    writeln(A:6, B:11, B/A:14:6);
    (**** Compute next Fibonacci number ****)
    S := A + B;
    A := B;
    B := S;
    C := C + 1;
  until C > T;
end.
```

ENTER NUMBER OF TERMS 19

TERM A	TERM B	RATIO A/B
1	1	1.000000
1	2	2.000000
2	3	1.500000
3	5	1.666667
5	8	1.600000
8	13	1.625000
13	21	1.615385
21	34	1.619048
34	55	1.617647
55	89	1.618182
89	144	1.617978
144	233	1.618056
233	377	1.618026
377	610	1.618037
610	987	1.618033
987	1597	1.618034
1597	2584	1.618034
2584	4181	1.618034
4181	6765	1.618034

6.5 FIBONACCI SORTING

In the computer field it is often necessary to sort a string of items, stored serially on magnetic tape, into ascending or descending order. The need arises mainly in commercial applications, and often the series contains many thousand items. Multi-stage sorting processes are essential to deal with such large quantities of data quickly. The basic principle consists of sorting quite small blocks, perhaps of 1000 items each. Then these are merged into sorted blocks of 2000 each and so on. The merging process continues until there is a single sorted block.

The following Fibonacci 3-magnetic tape deck sorting technique has often been used. The initial stage is designed to produce an F-number of blocks; we take 13 for illustrative purposes, but it could be much larger. They are initially stored on tape deck A; 5 of them are at once copied to tape deck B, while tape deck C is empty. The following list shows successive states of the 3 tape decks being used.

NUMBER THEORY TRIVIA

The twelfth Fibonacci number is the square of 12.

$$12^2 = 144$$

A mathematician at the University of London proved in 1963 that 144 is the only square number in the entire Fibonacci number sequence (except 1).

NUMBER THEORY TRIVIA

The sum of the numbers at any given point in the **Fibonacci number sequence** is not a Fibonacci number; but if you add 1 to it, the result is the number two terms ahead.

1+1+2+3+5+. . .+34+55 = 143

which is 144 minus 1.

Tape Deck A	Tape Deck B	Tape Deck C
8	5	0
3	0	5
0	3	2
2	1	0
1	0	1
0	1	0

The second state arises by merging all the blocks of deck B with the first 5 of deck A, putting results on the deck C. This leaves deck A with 3 blocks and deck B free. Therefore the next merge is on to deck B, 3 blocks coming from deck A, and leaving it free, while the number on deck C is reduced to 2. The culmination of the process should now be apparent from the table.

I'll take the microcomputer, Zula . . . you carry the reference manuals.

1	2	3
2		
3		

Find the unique solution to this crossnumber puzzle. The clues:

Across: 1. A Fibonacci number; 2. A square; 3. A perfect number.

Down: 1. A square; 2. A cube; 3. A multiple of a square.

The answer is shown on page 300.

REVIEW EXERCISES

1. Which of the following sequences are arithmetic? If the sequence is arithmetic, give the common difference and the next term.
   a. 41, 45, 49, 53, . . .
   b. 34, 51, 68, 84, . . .
   c. 17, 24, 31, 38, . . .
   d. 83, 102, 121, 142, . . .

2. Which of the following sequences are geometric. If the sequence is geometric, give the common ratio and the next term.
   a. 2, 6, 18, 54, . . .
   b. 6, 12, 24, 96, . . .
   c. 2, 12, 72, 432, . . .
   d. 3, 15, 45, 215, . . .

3. Fill in the missing term
   cat, three, giraffe, four, tiger, five, bear, ______ , . . .

4. Fill in the missing term
   A, two, B, four, C, eight, D, sixteen, F, __________ , . . .

5. What are Fibonacci numbers?

6. Write a short paper about Leonardo Fibonacci.

7. Write out the first 15 terms of the Fibonacci number sequence.

8. What is the thirty-first Fibonacci number?

9. Modify the Fibonacci Numbers program in Section 6.2 so that the Fibonacci numbers will be printed in horizontal rows.

10. The Fibonacci numbers are generated by letting the first two numbers of the sequence equal 1; from that point each number may be found by taking the sum of the previous two elements in the sequence. So you get 1, 1, 2, 3, 5, 8, 13, etc. Prepare two lists: one with the first ten and the other with the second ten. For each element from two to nineteen find the difference between the square of the element and the product of the elements immediately preceding and following. In other words, print $F(I)^2 - F(I - 1) \times F(I + 1)$.

11. Modify the Fibonacci Number program in Section 6.2 so that it will generate and print the first 50 Fibonacci numbers.

12. Compare the Fibonacci number sequence with the tetranacci number sequence.

13. Compare the Fibonacci sequence with the tribonacci number sequence.

14. Compare the Fibonacci sequence with the Lucas number sequence.

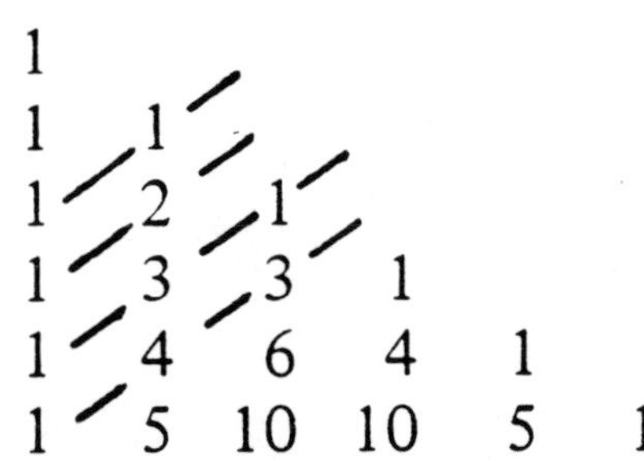

The Fibonacci numbers are closely related to the binomial coefficients. Look at the above Pascal's triangle. The sums of the numbers on the diagonals rising at 45° are Fibonacci numbers.

I wish we had a computer to work this out for us.

7

π IN THE COMPUTER AGE

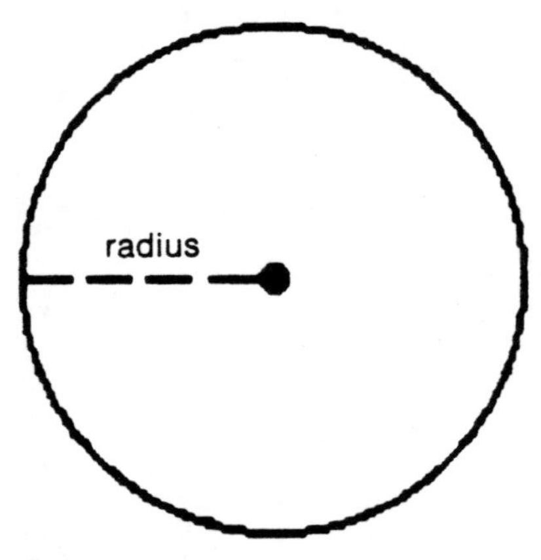

Circle
Circumference = $2\pi r$
Area = πr^2

Sphere
Volume = $4/3\pi r^3$
Surface area = $4\pi r^2$

PREVIEW

π is the ratio of the circumference to the diameter of a circle. Two of the most common formulas for using π(pi) are shown at the side. For some four thousand years, mathematicians have been finding better and better approximation for π. The ancient Egyptians used a method for finding the area of a circle that is equivalent to a value of 3.1605 for π. The Babylonians used numbers that give 3.125 for π. In the Bible, we find a verse describing a circular pool at King Solomon's temple, about 1000 B.C. The pool is said to be ten cubits across, "and a line of 30 cubits did compass it round about." This implies a value of 3 for π.

π, and the identification of its complete value, has intrigued mathematicians from the time of its discovery in early Greece to the present day. The value of

3.141592653589793 . . .

represents the beginning of an infinite series of non-repeating digits. π carried out to 200 decimal places is shown in Figure 7-1. The universally accepted symbol π was popularized by Leonhard Euler in the 1700s.

Everyone has heard the classroom story about how π calculated many times over would fill all the chalkboards with small print. In the past few years, large supercomputers have been used to compute π

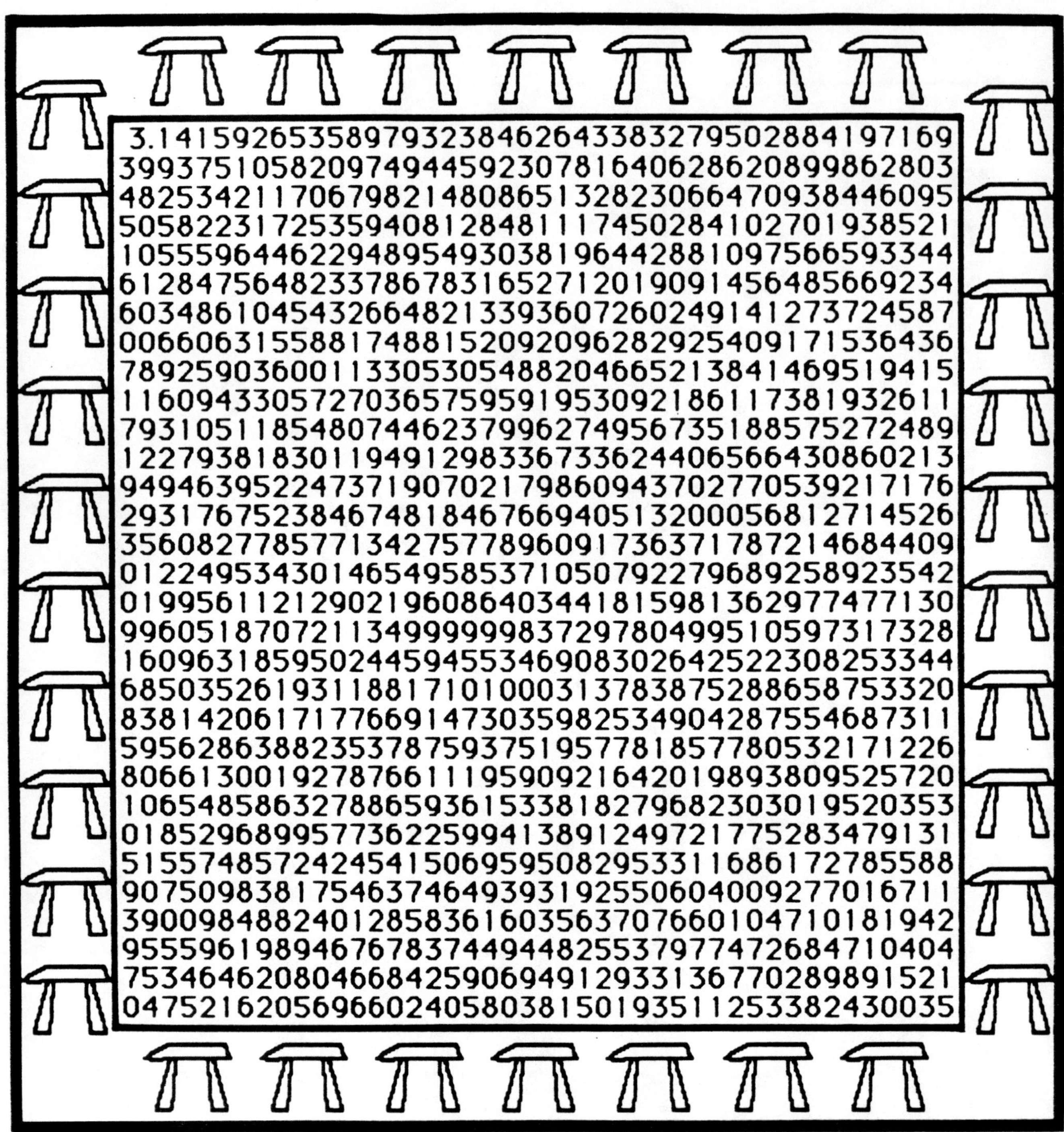

Figure 7-1
π carried out to 1362 decimal places.

to several million places. Microcomputers can also be used to compute approximations of π. However, better results will occur if one uses a microcomputer that uses double or triple precision arithmetic.

After you finish this chapter, you should be able to:

1. Identify a geometric method for approximating values of π.
2. Identify several infinite series that can be used to produce values of π.
3. Identify some of the computers that were used to compute π.

4. See how a microcomputer can be used to generate values of π.
5. Identify some of the number patterns that exist in large values of π.

Archimedes (287-212 B.C.) was a native of Syracuse, a Greek city on the island of Sicily, who made many important contributions to mathematics. One of the first times very large numbers were used was by Archimedes in about 250 B.C., when he reportedly computed the number of grains of sand in the universe to be 10^{63}. Archimedes was one of the greatest mathematicians of all times.

7.1 EARLY METHODS OF COMPUTING π

References to the value of π were made in earliest literature. The Rhind Papyrus from Egypt (dated to 1700 B.C.) used the value of $(16/9)^2$.

The first person to find a reliable way of computing π was the Greek mathematician and inventor Archimedes in about 240 B.C. Archimedes used the perimeters of inscribed and circumscribed polygons to approximate the circumference of a circle (see Figure 7-2). By increasing the number of sides of the polygons, he could squeeze the value of π between the two bounds. Archimedes best results showed π to be between 223/71 and 22/7.

Archimedes is featured on this Italian stamp.

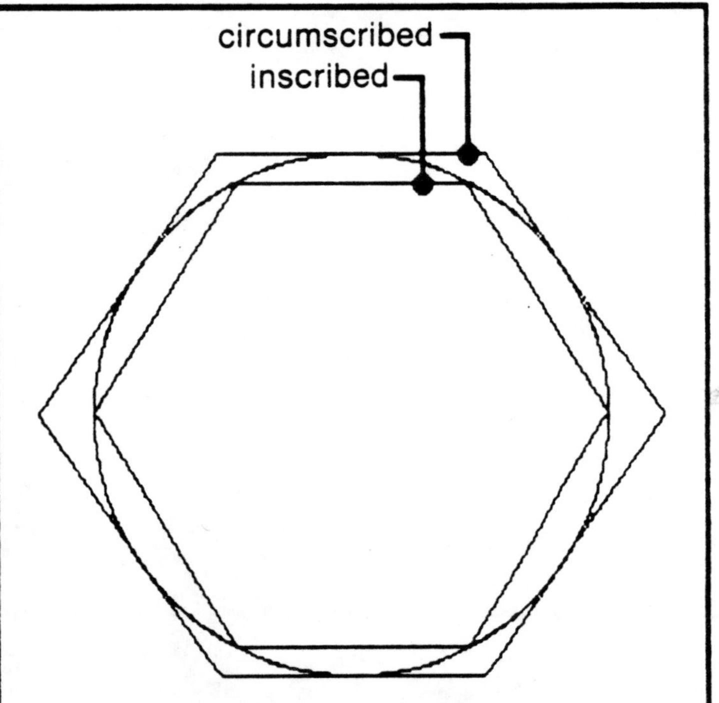

Inscribed polygons are those measured inside the circle (points touching). Circumscribed polygons are those measured outside the circle (flats touching). Archimedes used both polygon measurements more than 2,200 years ago to calculate π to within two decimal places of accuracy.

Figure 7-2
Polygon measurements

The ancients do not appear to have been successful or interested in computing π very accurately. The most precise value computed was 3.1416. This was worked out by the mathematician and astronomer Claudius Ptolemy about 150 A.D.

Tsu Ch'ung-chih (about 480 A.D.), the Chinese mathematician gave 22/7 as an "inaccurate value" of π, and 355/113 as the "accurate value," and he showed that π lies between our present decimal forms 3.1415926 and 3.1415927.

In 480 A.D., the Chinese astronomer Tsu Chung-chih calculated π as 3.1415929. This fraction of Chung-chih is about as close an approximation of π as is needed in most applications.

In 530, the early Hindu mathematician Aryabhata gave 3.1416 as an approximate value for π. It is not known how this result was obtained. In 1150, the later Hindu mathematician, Bhaskara, used the square root of 10 as a value for π.

Early methods of computing π were geometric in nature. As mathematical analysis techniques grew, the geometric method was abandoned in favor of numerical methods.

In the sixteenth century, several series of converging fractions were found and used to compute values of π. In 1579, the French mathematician Francois Viete developed the following formula and, by hand, found nine decimal places for π!

$$\pi = \frac{1}{\sqrt{\frac{1}{2}} \cdot \sqrt{\frac{1}{2} + \frac{1}{2}\sqrt{\frac{1}{2}}} \cdot \sqrt{\frac{1}{2} + \frac{1}{2}\sqrt{\frac{1}{2} + \frac{1}{2}\sqrt{\frac{1}{2}}}} \cdot \ldots}$$

Francois Viete (1540-1603) was one of the greatest French mathematicians of the 16th century. He was a lawyer at the court of Henri IV of France and studied equations. Viete simplified the notation of algebra and was among the first to use letters to represent numbers.

The greater the number of terms used in the computation, the closer you get to the exact value of π. Viete's work with π is significant in that it was the first time the exact value of π was expressed in a regular mathematical pattern. Note that the denominator is an infinite product of expressions of square roots.

Sir Isaac Newton (1642-1727), a physicist and mathematician, among many other great discoveries, invented the differential and integral calculus (Leibniz discovered it independently of Newton some 10 years later). Newton also found a way to calculate π. For a giant like Newton, the calculation of π was trivial, and indeed, in his *Method of Fluxions and Infinite Series*, he devotes only a paragraph of four lines to it, apologizing for such a triviality with a *by the way* in parentheses — and then gives its value to 16 decimal places.

In 1593, Adriaen van Roomen, more commonly referred to as Adrianus Romanus, of the Netherlands, found π correct to 15 decimal places by the classical method, using polygons having 2^{30} sides.

Around 1610, Ludolph van Ceulen of Germany computed π to 35 decimal places by the classical method of using polygons up to 2^{62} sides. He spent a large portion of his life on the task. This approximation was engraved on his tombstone, and even today, π is referred to in Germany as "the Ludolphine numbers." Here are the first twenty digits of his endeavor:

$$\pi = 3.14159\ 26535\ 89793\ 23846\ \ldots$$

In 1621, the Dutch physicist Willebrord Snell computed π to 35 decimal places. In 1630, Grienberger, computed π to 39 decimal places.

In 1650, the English mathematician, John Wallis, developed the series:

$$\pi = 2 \times \frac{2 \times 2 \times 4 \times 4 \times 6 \times 6 \times 8\ \ldots}{1 \times 3 \times 3 \times 5 \times 5 \times 7 \times 7\ \ldots}$$

In 1671, the Scottish mathematician James Gregory discovered an infinite series for the arc tangent. Recall that in a right triangle the tangent of the angle A is the ratio x = a/b, which we write as tan A = x = a/b. Usually, we know the angle A and we find the ratio x. However, if we start off knowing the ratio x and find the angle A, we find the arc tangent. So the arc tangent is just an angle, and the formula is written as arctan x = A (see Figure 7-3). Gregory's formula was

$$\arctan x = x - x^3/3 + x^5/5 - x^7/7 + \ldots$$

where x is equal to (or greater than) −1, and equal to (or less than) +1.

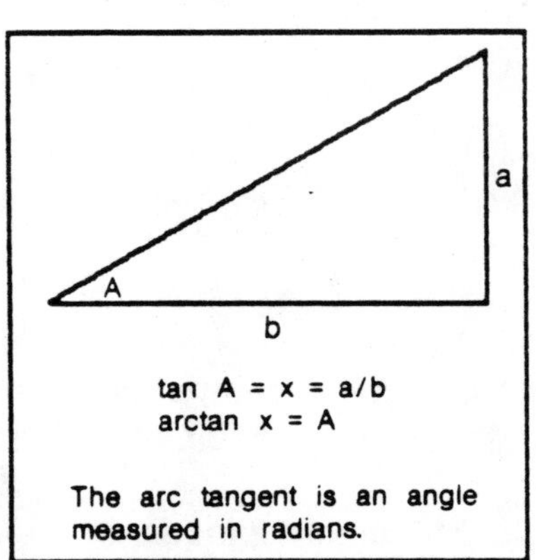

Figure 7-3
Arc tangent

The Swiss mathematician Leonhard Euler came up with an interesting formula for π:

$$\pi = 4(\arctan(1/2) + \arctan(1/3))$$

π can be expressed in terms of two arc tangents that, in turn, can be computed using Gregory's formula. Combining Gregory's and Euler's formulas yields an effective method for computing π to a large number of decimal places.

In 1674, the German mathematician, Gottfried Leibniz, discovered the following formula for π.

$$\pi = 4(1 - 1/3 + 1/5 - 1/7 + 1/9 - 1/11 + 1/13 - \ldots)$$

Gottfried Leibniz (1646-1716), a German mathematician and philosopher, thought it was unworthy of excellent men to lose hours like slaves in the labor of calculations which could safely be relegated to anyone else if machines were used. Leibniz made considerable contributions to mathematicians but is unfortunately often remembered most vividly for the controversy over the "invention" of the calculus. But it is his notation for the calculus, which he introduced in 1675, which has survived.

This formula presents π as the limit of an infinite series of fractions whose denominators are odd numbers and whose signs alternate. The simplicity of the Leibniz series makes it one of the best known expressions for π. However, it is only when the number of terms becomes extremely large that the series becomes a powerful tool for approximating π.

Many infinite series relating to π were discovered in the late 1600s and early 1700s. Here are a few which date from this era of mathematics:

$$\pi^2/6 = 1 + 1/1^2 + 1/2^2 + 1/3^2 + 1/4^2 + \ldots$$

$$\pi = 16(1/5 - 1/(3 \times 5^3) + 1/(5 \times 5^5) - 1/(7 \times 5^7) + \ldots) - 4(1/239 - 1/(3 \times 239^3) + 1/(5 \times 239^5) - \ldots)$$

$$\pi = 2\sqrt{3}(1 - 1/(3 \times 3) + 1/(5 \times 3^2) - 1/(7 \times 3^5) + \ldots)$$

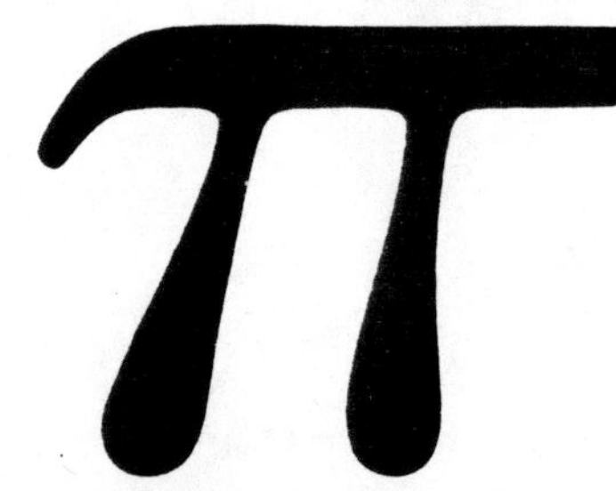

The symbol for the circle ratio was first used in 1706 by William Jones (1675-1749) who occasionally edited and translated (from Latin) some of Sir Isaac Newton's works, and who himself wrote on general mathematics. However, Jones did not have the weight to make his notation generally accepted. In 1739, Leonhard Euler used the notation and it became a standard symbol.

Some of these infinite series are well adapted for computation and through them π was obtained to a great many decimal places.

In 1699, Abraham Sharp computed π to 71 decimal places. In 1706 John Machin computed its value to 100 figures. In 1719, the French mathematician DeLagny obtained 112 correct places for π.

NUMBER THEORY TRIVIA

On October 15, 1974, Michael Poulteney memorized π correctly to 3025 decimals. It took twenty-five minutes to write these decimals out!

Prior to 1761, mathematicians thought that if π were taken to enough decimal places, a pattern might be observed. In 1761, Johann Heinrich Lambert proved that π is an irrational number. It is a nonrepeating, nonterminating decimal. Its exact value cannot be computed.

In 1794, Vega, a man who had computed a famous table of logarithms, calculated a value of π to 137 places. In 1841, an Englishman, William Rutherford, calculated π to 208 places, of which 152 were later found to be correct. In 1844, Zacharias Dase found π correct to 200 places using Gregory's series in connection with the relation

$$\pi = 4(\arctan(1/2) + \arctan(1/5) + \arctan(1/8))$$

In 1853, Rutherford returned to the problem and computed π to 400 correct decimal places.

In 1873, a British mathematician, named William Shanks, had worked out π to 707 decimals, and for more than seven decades no one bothered to check his figures. Poor Shanks! He had spent twenty years doing his calculations by hand only to fumble after 527 correct decimals. The 528th is 4, but Shanks called it a 5, and from there on his digits are wrong. The error went undetected until 1945, when another Englishman, D. F. Ferguson, discovered it. Four years later the value of π was accurately extended to 1120 decimals by two Americans, John Wrench Jr. and Levi Smith, in what turned out to be the last effort to compute π on a pre-electronic calculator.

In 1882, Ferdinand Lindemann proved that π is transcendental. That is, π cannot be the root of an algebraic equation, such as

$$\pi^4 - 240\pi^2 + 1492 = 0$$

or any other similar equation containing π as a base. Lindemann proved this by relating π to e, the base of the natural logarithms, which was already known to be transcendental.

After Lindemann's proof, the computation of π went to sleep again. The computation had been pushed to limits which made further work intolerably difficult and tedious, and the last vestige of a reason for additional computation had long disappeared. To go further one would need improved means for computing and a significant reason for doing so.

I'll never catch up to you.

4 π

Mathematics has curious powers of regeneration. In the late 1940's two new mathematical streams met and led to the renewed computation of π. The development of high speed computers gave the means for computation, and inquiries as to how the digits of π behaved statistically, provided the motive.

ENIAC, the first large scale electronic digital computer, was used to compute π to 2037 places.

π is an interesting enigma. While it is challenging to try to develop new infinite series to provide longer lists of π's decimals, no more than 10 decimal places are needed for even the most precise practical applications. It has been stated that 10 decimal places would describe the circumference of the planet earth with an accuracy within a fraction of an inch.

7.2 COMPUTER CALCULATIONS OF π

The first electronic computation of π was made in September 1949 on ENIAC (Electronic Numerical Integrator And Computer) at the Army Ballistic Research Laboratories in Aberdeen, Maryland; it calculated π to 2037 decimal places in 70 hours. The computer was programmed in accordance with Machin's formula in the form:

$$\pi = 16 \arctan(1/5) - 4 \arctan(1/239)$$

In 1955, NORC (Naval Ordnance Research Calculator) at Dahlgren, Virginia, was programmed to compute π to 3089 significant places; the run took only 13 minutes.

This record was broken in 1959 when an IBM 704 computer at the Paris Data Processing Center was programmed according to a combination of Machin's formula and the Gregory series; it yielded 10,000 decimal places in 1 hour and 40 minutes. The program was prepared by F. Genuys who did it as a training problem.

In England in the same year, at the Ferranti Computer Center, London, a Ferranti PEGASUS computer computed π to 10,021 decimal places in 33 hours.

Also in 1959, an IBM 704 computer at the Commissariat a l'Energic Atomique in Paris, France, and 16,167 places were obtained in a little over four hours. Machin's formula was also the basis of a program run on an IBM 7090 computer at the London Data Centre in 1961, which resulted in 20,000 decimal places and required only 39 minutes running time.

By this time, the limit of the then available computer memories had almost been reached. Further substantial increases in the number of decimal places could have been obtained only by modifying the programs to use more computer time and therefore to run into unreasonable costs.

In 1961, Daniel Shanks and John Wrench increased the speed of the computation by a factor of about 20. In part, this was due to a faster computer (an IBM 7090 at the IBM Data Processing Center, New York), but they also used several tricks in programming it; in particular, they abandoned Machin's formula in favor of the formula:

$$\pi = 24 \arctan(1/8) + 8 \arctan(1/57) + 4 \arctan(1/239)$$

which was found by Stormer in 1896. The run resulted in 100,265 places; and took 8 hours and 43 minutes. A computation of this kind involves billions of individual arithmetic operations, and if a single one of these is mistaken, the entire subsequent operation may yield an erroneous result. It is therefore necessary to check the result. For this, Shanks and Wrench used a special method which calculates π by a different formula, but uses the partial results of the original run in such a way that the check takes less time than the original computation.

In 1966, two French mathematicians, Jean Gilloud and J. Fillatoire used an IBM 7030 (STRETCH) computer to compute 250,000 decimals of π.

Just one year later, in 1967, Jean Gilloud and Michele Dichampt computed π to 500,000 decimal places in 28 hours and 10 minutes using a Control Data CDC 6600 computer system. It took them an additional 16 hours and 35 minutes to check the result. The program was again based on Stormer's formula and the Shanks and Wrench method for checking the digits.

In 1976, Gilloud and Martine Bouyer computed π to a million decimals, a calculation that took 23 hours and 18 minutes on a CDC 7600 computer. The French atomic energy commission considered the results important enough to publish a 400-page book.

Gilloud devised a new computer method to make his 1976 calculation, and three years later, Eugene Salamin invented a more streamlined algorithm. Salamin had planned to use his method to calculate π to 10 million places, however, his project was cancelled due to being unable to obtain sufficient computer time. Salamin's algorithm is based on an infinite series of fractions that when extended converge with great rapidity on π. The number of calculated digits doubles at each step. At first, Salamin thought his series was original. Then he learned he had rediscovered a formula published in 1818 by the German mathematical genius, Karl Gauss. No one had considered using it for computing π because it involved such time-consuming multiplication. Only with the advent of high-speed supercomputers and clever new algorithms for multiplying has it become practical to put Salamin's (or Gauss') algorithm to work calculating π.

Is a million digits the record? Not by a long shot! In 1981, Kazunori Miyoshi and Kazuhika Nakayama used a FACOM M-200 computer to compute π to 2,000,038 decimal places. This computation was performed at the University of Tsukuba using the formula of Klingenstierna.

$$\pi = 32 \arctan(1/10) - 4 \arctan(1/239) - 16 \arctan(1/515)$$

It is one of the family of arctangent formulas that have been traditionally used for such computations.

LARGEST VALUE
OF
π

- 480 MILLION DIGITS
- DISCOVERED BY
DAVID CHUDNOVSKY and
GREGORY CHUDNOVSKY

In 1982, the Japanese adopted Salamin's approach to the challenge. Yoshiaki Tamura and Yasumasa Kanada of the University of Tokyo, using the HITAC M-280H supercomputer, calculated π to 4,194,293 decimals. This computation required 2 hours and 53 minutes of computer time.

A year later, in 1983, Tamura and Kanada calculated π to 2^{24}, or 16,777,216, places in less than thirty hours. In 1984, these results were verified on an even faster computer, to 10,013,395 decimals. In 1987, Kanada used a NEC SX-2 supercomputer to approximate π to 134,217,728 decimal places. In 1988, Kanada computed π to 201,326,000 places. In 1989, two Columbia University research scientist brothers, David and Gregory Chudnovsky, computed π to 480 million decimal places. This number would stretch for 600 miles if printed.

Does such precision have any conceivable use? Probably not. While it is doubtful that there is any repetitive pattern in the sequence of digits, some mathematicians, lacking formal proof that no such pattern exists, may examine the new value of π in hopes of finding one. Perhaps the real object of such an exercise is more philosophical than practical.

William James wrote in 1909: "The thousandth decimal of π sleeps there [in the world of abstraction] though no one may ever try to compute it." James might be surprised to learn that the thousandth decimal has been far exceeded. (Incidentally, the thousandth decimal of π is 9; the millionth decimal is 1; the two-millionth is 9.) But he could still ponder that ghostly, remote billionth digit.

Figure 7-4 shows some of the key milestones in the computation of π.

7.3 IS π PATTERNLESS?

The more digits of π that are known, the more mathematicians and computer scientists hope to answer a major unsolved problem of number theory: Is π's sequence of digits totally patternless, or does it exhibit a persistent, if subtle, deviation from randomness?

To explain a random sequence of digits, consider this analogy: Imagine yourself at a casino gambling table betting on the next digit to appear while a sequence of digits is being generated, perhaps by a roulette wheel. If there is no possible way of predicting the next digit with a probability of better than one in ten, then the sequence is random. In this sense, π is certainly not random. You can always make your own calculation and predict the next digit with certainty. In another sense, however, π can be called random. As far as anyone knows, it shows no trace of a pattern, no kind of order, in the overall arrangement of its digits. Every digit has the same probability (one in ten) of appearing at any one spot, and the

π

	DECIMAL PLACES
Archimedes, 240 B.C.	2
Ptolemy, 150 A.D.	4
Chung-Chih, 450 A.D.	7
Viete, 1579	10
Romanus, 1593	16
van Ceulen, 1610	35
Sharp, 1699	72
Machin, 1706	101
Vega, 1794	137
Dase, 1844	201
Rutherford, 1853	441
Shanks, 1873	527
Wrench and Smith, 1949	1120
ENIAC, 1949	2037
NORC, 1954	3093
PEGASUS, 1957	10,021
Genuys, (IBM 704), 1958	10,000
Genuys, (IBM 704), 1959	16,167
Shanks and Wrench, (IBM 7090), 1961	100,265
Gilloud and Fillatoire, (IBM 7030), 1966	250,000
Gilloud and Dichampt, (CDC 6600), 1967	500,000
Gilloud and Bouyer, (CDC 7600), 1976	1,000,000
Miyoshi and Nakayama, (FACOM M-200), 1981	2,000,038
Tamura and Kanada, (HITAC M-280H), 1982	4,194,293
Tamura and Kanada, (HITAC M-280H),1983	8,388,608
Tamura and Kanada, (HITAC M-280H), 1983	16,777,216
Bailey, Borwein, Borwein, (Cray 2), 1986	29,360,000
Kanada, (NEC SX-2), 1987	134,217,728
Kanada, (Hitachi supercomputer), 1988	201,326,000
Chudnovsky, David and Gregory, 1989	480,000,000

Figure 7-4
Key milestones in the computation of π.

NUMBER THEORY TRIVIA

The following equation has a root that gives a close approximation to π.

$$19x^2 - 39x - 65 = 0$$

same conformity to randomness applies to so-called doublets, triplets, or any specified pattern of digits, adjacent or separated, in π's endless stream of digits. Herein lies the rub: no one has proved that π is patternless, nor has anyone proved it isn't.

As π is extended to an ever-larger number of decimal places, the digits 0 through 9 occur with about equal regularity. Although these digits could be distributed in a large number of ways, statistical tests show that they seem to be distributed randomly. Listed below are the frequencies of occurrence of the ten digits in the first ten thousand decimal places of π:

zero	969
one	1,026
two	1,021
three	974
four	1,012
five	1,046
six	1,021
seven	970
eight	948
nine	1,014

Yet assuming π is patternless, it doesn't follow that π doesn't contain an endless variety of remarkable patterns that are the result of pure chance. For example, starting with π's 710,100th decimal is the stutter 3333333. Another run of 3s starts with the 3,204,765th decimal. There are runs of the same length, among the first ten million decimals of π, of every digit except 2 and 4. Digit 9 leads with four such runs; 3, 5, 7, and 8 each have two seven-runs; and 0, 1, and 6 have one run apiece. There are 87 runs of just six repetitions of the same digit, of which 999999 is the most surprising because it comes so soon, relatively speaking: it starts with the 762nd decimal.

The ascending sequence 23456789 begins with decimal 995,998, and the descending sequence 876543210 starts at decimal 2,747,956. Among the first ten million decimals of π, the sequence 314159 - the first six digits of π appears no fewer than six times. The first six digits of e, a celebrated number in mathematics that can be defined as the basis of natural logarithms, occurs eight times, not counting one appearance (at decimal 1,526,800) of 2718281, the first seven digits of e. Even more unexpected is the appearance (starting with the 52,638th decimal) of 14142135, the first eight digits of the square root of 2.

We can look for still other oddities in π. If the first n digits of π form a prime number, let's call it a pifor (π forward) prime. Only four such numbers are known:

3
31
314159
3 14159 26535 89793 23846 26433 83279 50288 41

The fourth was proved to be a prime in 1959 by Robert Baillie and Marvin Wunderlich of the University of Illinois. Is there a fifth? Probably, but it may be some time before anyone knows.

What about piback primes, or the first n digits of π running backward? Six pibacks can be easily identified:

3
13
51413
951413
2951413
53562951413

Note that all pibacks end in 3. Sharp-eyed readers may also note that the first three pibacks are in fact reversals of three pifors.

7.4 ARCHIMEDES POLYGON PROGRAM

As discussed in Section 7.1, Archimedes was interested in computing π and he succeeded in coming closer to its true value than any of the mathematicians of ancient Egypt or Babylonia. He did this by using inscribed and circumscribed regular polygons to get more and more accurate results. The method worked like this: A circle having a diameter of 1 unit length has an equilateral triangle (a triangle is a three-sided polygon) inscribed in it. Also, an equilateral triangle is circumscribed about the circle. Clearly, the circumference of the circle is less than the perimeter of the large triangle and more than that of the small triangle. Then, the circumference of the circle must equal π since its diameter equals 1 and **circumference** = π × **diameter**. Therefore, we can say that π has a value between the value of the perimeter of an inscribed triangle and that of the circumscribed triangle.

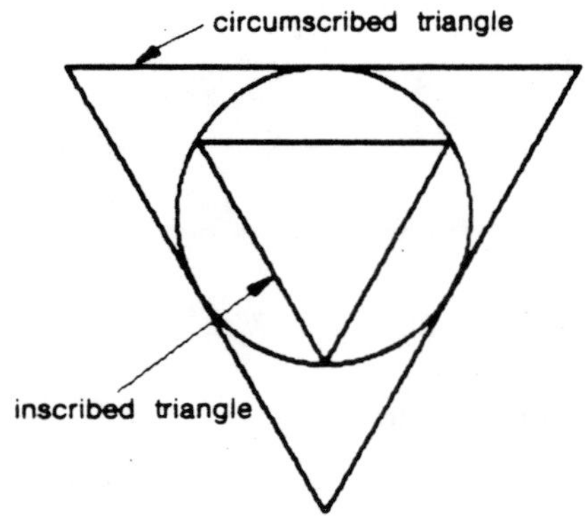

By using polygons with ever-increasing numbers of sides, it is possible to "close in," more and more, on the exact value of the circumference of the circle. That value is equal to π. As the number of sides increases, the more accurate becomes the estimated value for π.

Consider a polygon circumscribed around a circle of radius 1. The perimeter of a polygon equals the length of one side times the number of sides. Since the tangent of x = BC/AC, but AC = 1, then

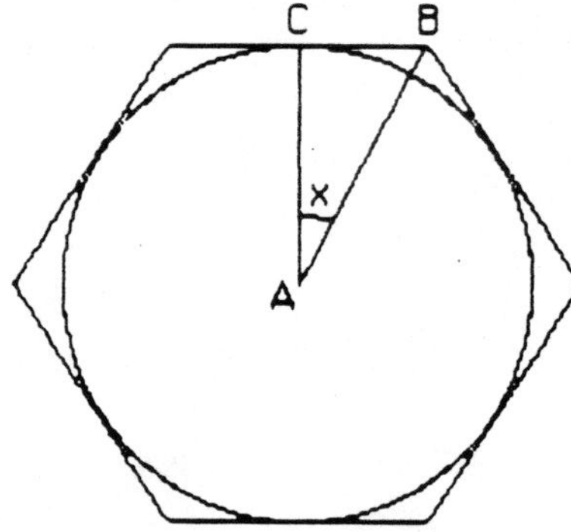

tan(x) = BC and the length of a side = 2 tan(x). Since the circumference = $2\pi r$ and r = 1, then π is the circumference (or perimeter of an n-sided polygon) divided by 2. The perimeter of an inscribed polygon is equal to the number of sides times sine(x) = cosine(x).

The following program converges to the value of π using the Archimedes polygon method. Unfortunately, there is one large flaw in this program, because degrees must be converted into radians. This means, of course, that you must already know the value of π, since the conversion factor used is

$$360 \text{ degrees}/2\pi = 57.29578 \text{ radians}$$

Ignoring this flaw, it is interesting to see how Archimedes method converges on the value of π. Remember that this program converges geometrically rather than arithmetically. A good exercise would be to figure out a degrees-to-radians conversion factor that does not require that you know the value of π.

```
program ArchimedesPolygon;
var
  L      :Integer;(* Looping index *)
  N      :Integer;(* Number of sides of polygon *)
  C      :Real;(* Circumscribed polygon *)
  I      :Real;(* Inscribed polygon *)
  X      :Real;(Angle in radians *)
begin
  (**** Print heading for table ****)
  writeln('INSCRIBED               CIRCUMSCRIBED');
  writeln(' POLYGON                     POLYGON');
  writeln(-----------------------------------------');
  (**** Compute approximation for PI ****)
  N := 6;
 repeat
    N := N + 200;
    (**** Convert degrees to radians ****)
    X := 360.0 / (N * 57.29578);
    (**** Compute perimeter of inscribed polygon ****)
    I := (N * SIN(X) * COS(X)) / 2.0;
    (**** Compute perimeter of circumscribed polygon ****)
    C := (N * SIN(X) / COS(X)) / 2.0;
    (**** Print approximations of PI ****)
    writeln(I:10:9,'              ',C:10:9);
  until N > 10000;
end.
```

COMPILE/RUN

```
INSCRIBED          CIRCUMSCRIBED
 POLYGON              POLYGON
-----------------------------------------
3.139644563         3.142567203
```

3.141091041	3.141843456
3.141367481	3.141705207
3.141465352	3.141656267
3.141510927	3.141633478
3.141535778	3.141621052
3.141550801	3.141613540
3.141560570	3.141608656
3.141567277	3.141605302
3.141572080	3.141602901
3.141575636	3.141601122
3.141578344	3.141599769
3.141580452	3.141598714
3.141482126	3.141597878
3.141583476	3.141597202

7.5 LEIBNIZ SERIES PROGRAM

As discussed in Section 7.1, Gottfried Leibniz found an infinite series that can be used to grind out increasingly accurate values for π:

$$\pi = 4(1 - 1/3 + 1/5 - 1/7 + 1/9 - \ldots)$$

This series is called an arithmetic series and converges very slowly. The program shown below displays only every 500th value of the series. Don't be concerned if the program does not seem to be running very fast; a considerable amount of computing is taking place between each value that is printed.

```
program LeibnizSeries;
var
  I       :Integer;(* Value of denominator *)
  L       :Integer;(* Limit *)
  Q       :Integer;(* Display control character *)
  S       :Integer;(* Alternate sign of term *)
  P       :Real;(* Value of series *)
begin
  (**** Initialize variables ****)
  S := 1;
  I := 1;
  L := 1;
  Q := 0;
  P := 0;
  repeat
    repeat
      (**** Increase display control character ****)
      Q := Q + 1;
      (**** Compute value of series ****)
      P := P + S/I;
      (**** Increase value of denominator ****)
      I := I + 2;
      (**** Alternate sign of term ****)
      S := -S;
```

NUMBER THEORY TRIVIA

For it is unworthy of excellent men to lose hours like slaves in the labor of calculation which could safely be relegated to anyone else if machines were used.

Gottfried Leibniz
(1646-1716)
German mathematician
and philosopher

```
      (**** Is it time to display a value of PI? ****)
    until Q > 500;
    (**** Reset display control character ****)
    Q := 0;
    (**** Print value of PI ****)
    writeln(4.0 * P:10:9);
    L := L + 1;
  until L > 20;
end.
```

COMPILE/RUN

```
3.143588660
3.140594650
3.142257989
3.141093652
3.141991855
3.141259985
3.141877797
3.141343153
3.141814432
3.141393053
3.141774109
3.141426319
3.141746193
3.141450081
3.141725721
3.141467903
3.141710066
3.141481764
3.141697707
3.141492853
```

7.6 WALLIS SERIES PROGRAM

The series by John Wallis provides an interesting method for approximating π:

$$\pi = 2 \times 2/1 \times 2/3 \times 4/3 \times 4/5 \times 6/5 \times 6/7 \times 8/7 \times 8/9 \times \ldots$$

The Wallis Series program was run for the number of terms varying from 500 to 10,000 producing the following results.

NUMBER OF TERMS IN SERIES	APPROXIMATION OF π
500	3.140023814
1000	3.140807737
3000	3.141330882
5000	3.141435549
10000	3.141514029

As you see from the above data, this series is slow in generating an approximation for π.

John Wallis (1616-1703) devised a famous formula for generating π. Like Francois Viete, Wallis had found π in the form of an infinite product, but he was the first in history whose infinite sequence involved only rational operations; there were no square roots to obstruct the numerical calculation as was the case for Viete's formula.

```
program WallisSeries;
var
  A      :array[1 . . 10000] of Real;
  T      :Integer;(* Number of terms *)
  J      :Integer;(* Term value *)
  K      :Integer;(* Term value *)
  I      :Integer;(* Looping counter *)
```

```
  B       :Real;
begin
  write('ENTER SIZE OF SAMPLE TO BE RUN ');
  readln(T);
  (**** Initialize variables ****)
    J := 1;
    K := 2;
    B := 2.0;
  (**** Calculate terms and store in array ****)
    for I := 1 to T do
      begin
        A[I] := (K / J) * (K / (J + 2));
        K := K + 2;
        J := J + 2;
      end;
  (**** Multiply terms to obtain value for PI ****)
    for I := 1 to T do
      B := B * A[I];
  (**** Print PI ****)
    writeln('WALLIS APPROXIMATION OF PI = ',B:10:9);
end.
```

7.7 CONVERGING FRACTIONS PROGRAM

Another interesting series for computing an approximation for π is:

$$\pi = 6 \times \sqrt{1/3} \times (1 - 1/3 \times 3 + 1/3 \times 3 \times 5 - 1/3 \times 3 \times 3 \times 7 + 1/3 \times 3 \times 3 \times 3 \times 9 - \ldots)$$

The Converging Fractions program was run for several terms thus producing the following sample data.

NUMBER OF TERMS IN SERIES	APPROXIMATION OF π
5	3.141309
10	3.141593
15	3.141593
20	3.141593
25	3.141593
30	3.141593
35	3.141593

```
program ConvergingFractions;
var
  T       :Integer;(* Number of Terms *)
  P       :Integer;(* Power of 3 *)
  X       :Integer;(* 3-5-7-9-11-13. . . *)
  I       :Integer;(* Looping index *)
  S       :Integer;(* Sign change indicator *)
  C       :Real;(* Accumulation of terms *)
  A       :Real;(* Approximation of PI *)
function POWER(X:Real; P:Integer):Real;
```

NUMBER THEORY TRIVIA

The cube root of 31 is 3.1413806. This is a close approximation to π. The 9th root of 29,809 provides a better approximation of π, namely 3.14159149.

```
(**** This function returns the Pth power of X ****)
var
  Result      :Real;
  Index       :Integer;
begin
  Result := 1;
  for Index := 1 to P do
    Result := Result * X;
  POWER := Result;
end; (* end of POWER function *)
begin
  write('ENTER NUMBER OF TERMS ');
  readln(T);
  (**** Initialize variables ****)
  C := 1;
  P := 1;
  X := 3;
  S := -1;
  (**** Compute PI for T Terms ****)
  for I := 1 to T do
    begin
      C := C + 1 / (POWER(3,P) * X ) * S;
      (* Increase power of 3 *)
      P := P + 1;
      (* Increase term value *)
      X := X + 2;
      (* Change sign of term *)
      S := -S;
    end;
  (**** Print approximation of PI ****)
  A := 6 * SQRT(1/3) * C;
  writeln(' APPROXIMATION OF PI = ',A:8:6);
end.
```

7.8 SERIES OF FRACTIONS PROGRAM

The series

$$\pi = 4 \times (1/(2 \times 3 \times 4) - 1/(4 \times 5 \times 6) + 1/(6 \times 7 \times 8) - 1/(8 \times 9 \times 10) + \ldots) + 3$$

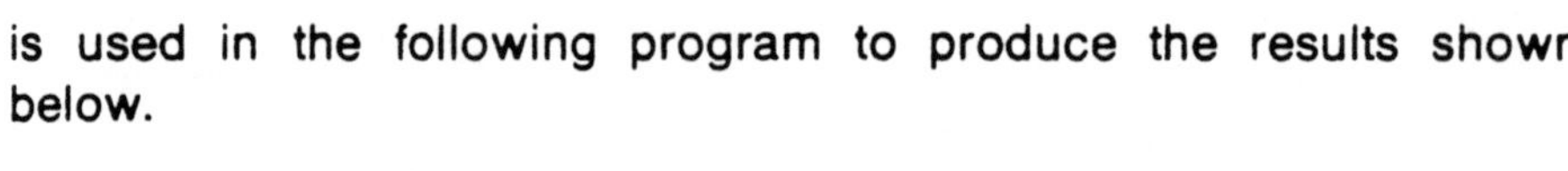

is used in the following program to produce the results shown below.

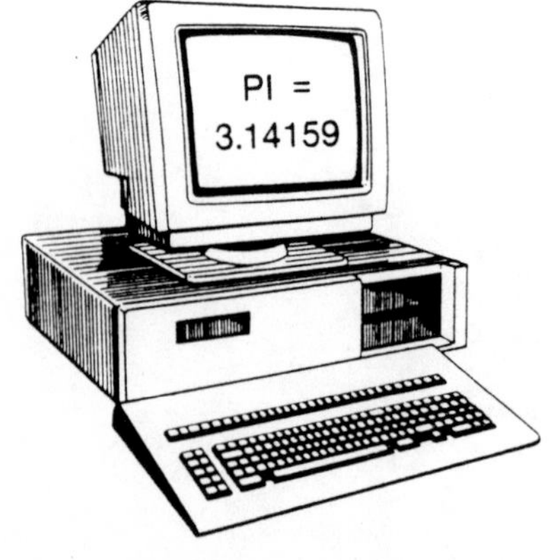

NUMBER OF TERMS IN SERIES	APPROXIMATION OF PI
20	3.141565735
40	3.141589029
60	3.141591553
80	3.141592183
100	3.141592411
120	3.141592512

140	3.141592564
160	3.141592594
180	3.141592611
200	3.141592623

This series started producing accurate approximations of π after only 40 terms.

```
program SeriesOfFractions;
var
  I,S,T          :Integer;
  A,B,C,P        :Real;
begin
  write('ENTER NUMBER OF TERMS: ');
  readln(T);
  writeln;
  (**** Initialize variables ****)
  P := 0.0;
  A := 2.0;
  B := 3.0;
  C := 4.0
  S := 4.0;
  for I := 1 to T do
    begin
      (**** Compute value of terms ****)
      P := P + 1.0 / (A * B * C) * S;
      (**** Increase A-B-C values by 2 ****)
      A := A + 2.0;
      B := B + 2.0;
      C := C + 2.0;
      (**** Change sign of term ****)
      S := -S;
    end;
  (**** Print approximation for PI ****)
  writeln('NUMBER OF TERMS IN SERIES = ',T);
  writeln('APPROXIMATION OF PI = ',((4.0 * P) + 3.0):10:9);
end.
```

REVIEW EXERCISES

1. Briefly define π.
2. Most calculations can be performed using a value of π with ______________ decimal places.
3. What is the name of the mathematician that first used the Greek letter π to represent the ratio of a circle's circumference to its diameter?
4. Who was the first person to find a reliable way of computing π?
5. In 240 B.C., how did Archimedes approximate the value of π?
6. Archimedes knew that π was less than ____________ and greater than 223/71.

NUMBER THEORY TRIVIA

The polynomial

$$11x^5 - 17x^4 - 42x^3 - 37x^2 - 27x + 42 = 0$$

has a root that gives a good approximation of π:

3.1415926535897

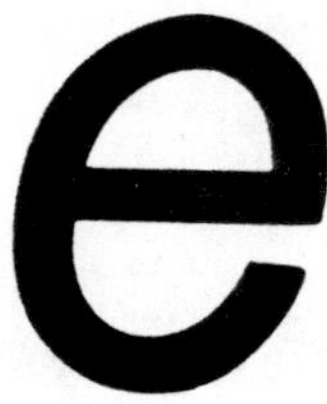

What is this number e? How big is it? Is it rational? It can be shown that e is an irrational number and it is, next to π, the most commonly used irrational number in all of mathematics.

e = 2.71828182845904523536O287 . . .

The following infinite series for e, which may well be the most important series expansion in the whole of mathematics, was discovered by Sir Isaac Newton and first appeared in a famous treatise written in 1665.

e = 1+1/1!+1/2!+1/3!+1/4!+ . . .

This series is simple and coverges very rapidly. The addition of twelve terms will produce an accurate value of e.

2.718281828

7. In the fifth century A. D., the Chinese astronomer Chung-Chih found an interesting fractional approximation of π. What was it?
8. What is the name of the French mathematician that represented π as an infinite product of square root expressions?
9. Most of the early methods used to approximate a value of π involved the use of ______________ ______________ .
10. In 1610, the German mathematician, Ludolph van Ceulen computed π to ______________ decimal places.
11. Identify the series that the English mathematician John Wallis used to approximate π.
12. What Scottish mathematician used an infinite series for the arc tangent to approximate π?
13. What formula did the Swiss mathematician Leonhard Euler use for π?
14. What expression did the German mathematician Gottfried Leibniz use for π?
15. What is Leibniz's approximation for π for six terms of the series?
16. For 11 terms?
17. In 1873, a British mathematician William Shanks worked out π to 707 decimal places. For more than seven decades no one bothered to check his figures. However, in 1945, another Englishman found that Shanks had made an error in the 528th decimal. Who made this discovery?
18. What is the name of the first electronic computer used to compute π?
19. In 1958, an IBM 704 computer was used to compute π to ______________ decimal places.
20. Who were the first people to compute π to more than 100,000 places?
21. In what year did Gilloud and Dechampt compute π to 500,000 decimal places?
22. Who were the French mathematicians who first computed π to a million places?
23. In 1983, ___________ ___________ and ______________ ______________, of the University of Tokyo, calculated π to more than 16 million places.
24. What type of computer was used to compute the largest value of π?
25. How long did it take the HITAC M-280H computer to calculate π to over 4 million digits?

26. The program used by Tamura and Kanada was based on an algorithm devised a decade ago by ________________________ ______________ .

27. There are four reasons why people use computers to find new approximations for π. Which one of the following reasons is incorrect?
    a. The calculation of π provides useful exercises for testing new computers.
    b. The more digits of π that are known, the more mathematicians hope to answer a major unsolved problem of number theory: Is π's sequence of digits totally patternless?
    c. Computing π on microcomputers has become one of the most important business applications for training new programmers.
    d. Calculations for π have useful spinoffs. Much is learned about calculating and checking large numbers on computers.

28. Modify the Archimedes polygon program so that it will compute π using a larger number of polygons.

29. Modify the Leibniz infinite series program so that it will print out every 1000th value of the series.

30. What result would be produced by the Wallis series program if the series contained 25,000 terms?

31. Write a program to approximate π using the following formula:

$$\pi^2/4 = 1 + 1/2^2 + 1/3^2 + 1/4^2 + \ldots$$

32. In 1914, the famous Indian mathematician Srinivasa Ramanujan, discovered a formula that produces π correct to eight decimal places.

$$\pi^4 = 2143/22$$

    Write a program that will produce a value for π using this formula.

33. The square root of 10 is π correct to one decimal. The square root of 2 plus the square root of 3 is correct to two decimals. The cube root of 31 is correct to three decimals. The square root of 9.87 (note the reverse counting order) is correct to a rounded four decimals. The square root of 146 times 13/50 is correct to a rounded six decimals. Write a program that will compute and display these results.

34. What supercomputer was used to compute π to 134,217,728 decimal places?

35. Who was the Japanese mathematician who first computed π to over 201 million decimal places?

NUMBER THEORY TRIVIA

The Russian mathematician Chebyshev (1821-1894) proved that between every integer (greater than 1) and its double, there is at least one prime, e.g., between 2 and its double, 4, is the prime number 3, between 4 and 8, the prime 7, and so on. Test his theorem by working out a dozen other exercises.

8
1
6
3
5
7
4
9
2

8 MAGIC SQUARES

PREVIEW

Magic squares and their properties have fascinated people for centuries. A novel approach to their creation is to generate them by a microcomputer. Those who are challenged by magic squares are always searching for new ways to generate them. Writing a microcomputer program to generate magic squares is not only an excellent problem-solving venture, but it also provides an alternate way of producing these interesting mathematical diversions.

After you complete this chapter, you should be able to:

1. Recognize several different types of magic squares.
2. Hand calculate magic squares.
3. Write programs that will generate magic squares.
4. Identify heterosquares and talisman squares.

8.1 INTRODUCTION

The study of magic squares probably dates back to prehistoric times. The first known example of a magic square is said to have been found on the back of a tortoise by the Chinese Emperor Yu in about 2200 B. C. This was called the lo-shu and appeared as a 3 by 3 array of numerals identified by knots in strings (see Figure 8-1). Black knots were used for even numbers and white ones for odd numbers. The sum along any row, column, or main diagonal is 15.

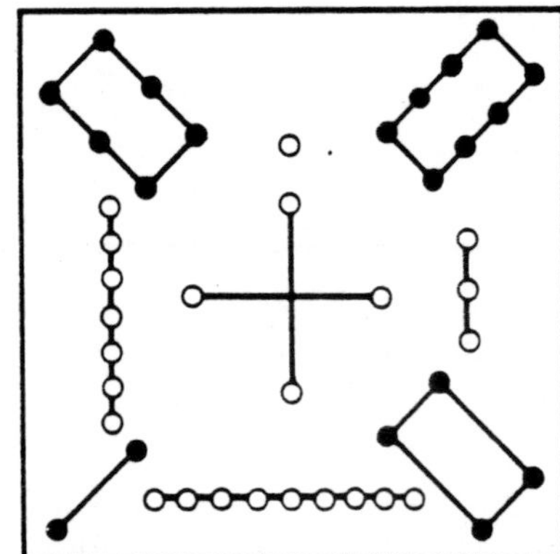

Figure 8-1
The first known example of a magic square comes from China. Legend tells us that around the year 2000 B.C., the Emperor Yu, of the Shang dynasty, received the above magic square etched on the back of a tortoise's shell. The incident supposedly took place along the Lo River, so this magic square has come to be known as the Lo-shu magic square. Can you translate this magic square into modern symbols?

It is recorded that as early as the ninth century magic squares were used by Arabian astrologers in their calculations of horoscopes. Hence the probable origin of the term "magic" which has survived to the present day.

Figure 8-2
Albrecht Durer's famous engraving *Melencolia* is not only a convincing testimonial for this side of his work, but also appears to be a song of praise to the beauty of mathematics. The engraving is preoccupied with things mathematical, geometry as well as number theory.

Figure 8-3
The 4 by 4 magic square shown here comes from the upper right-hand corner of the Melencolia engraving. This is a special magic square that has some special properties. A few are listed here. The corner numbers add up to the magic number 34. Opposite pairs of squares (3, 2, 15, 14) add up to 34. Slanting squares (2, 8, 15, 9) add up to 34. The sum of the first eight numbers equal the sum of the second eight numbers.

It was largely through the writings of Emanuel Moschopulus, a Greek mathematician and grammarian, who lived in Constantinople about 1300 A. D., that the knowledge of magic squares eventually spread throughout the western world and fascinated the great minds of the world.

One of the world's most famous engravings, Melencolia by Albrecht Durer (see Figure 8-2), depicts a 4 by 4 magic square (see Figure 8-3). The middle numbers in the last line represent the year 1514, in which Durer's print was made. He probably started from these two numbers and found the remaining ones by trial and error.

Pythagoras, in the fifth century, said that numbers govern the origin of all things and that the law of numbers is the key that unlocks the secrets of the universe. Magic squares brilliantly reveal the intrinsic harmony and symmetry of numbers; with their curious and mystic charm they appear to betray some hidden intelligence that governs the cosmic order that dominates all existence. Today, magic squares have lost some of their charm, however, they have become a popular mathematical diversion for mathematicians and computer scientists alike.

Magic squares have been used as magic charms in India, as well as being found on vases, fortune bowls, fans and other objects. Even today, they are widespread in Tibet and in other parts of the Far East. Shown above is an ancient Tibetan seal. A 3 by 3 magic square is found in the center of this seal.

Magic Squares are arrays of numbers such that the sum of the numbers in each row, column, or long diagonal is the same. The simplest magic square is one containing only nine boxes, with numbers from 1 to 9 inclusive, as shown in Figure 8-4. You will see that each row and each column adds up to 15. Also, the two diagonals add up to 15.

8	1	6
3	5	7
4	9	2

Figure 8-4
The simplest magic square containing the numbers from 1 to 9.

A magic square of order n is defined as n × n array of integers from 1 to n^2 such that the sum of each row, column, and main diagonal is a constant; that is, all the sums are the same. Figure 8-5 shows a magic square of order 4.

1	12	7	14
8	13	2	11
10	3	16	5
15	6	9	4

SUM OF ROW = 34
SUM OF COLUMN = 34
SUM OF MAIN DIAGONAL = 34

Figure 8-5
An order 4 magic square.

Benjamin Franklin was interested in magic squares. He often amused himself while in the Pennsylvania Assembly with magic squares. He also thought that mathematical demonstrations are better than academic logic for training the mind to reason with exactness and distinguish truth from falsity, even outside mathematics.

52	61	4	13	20	29	36	45
14	3	62	51	46	35	30	19
53	60	5	12	21	28	37	44
11	6	59	54	43	38	27	22
55	58	7	10	23	26	39	42
9	8	57	56	41	40	25	24
50	63	2	15	18	31	34	47
16	1	64	49	48	33	32	17

This 8 by 8 magic square was created by Benjamin Franklin. This square, whose magic constant is 260, contains all the numbers from 1 to 64. It also contains several other interesting properties. Can you find them?

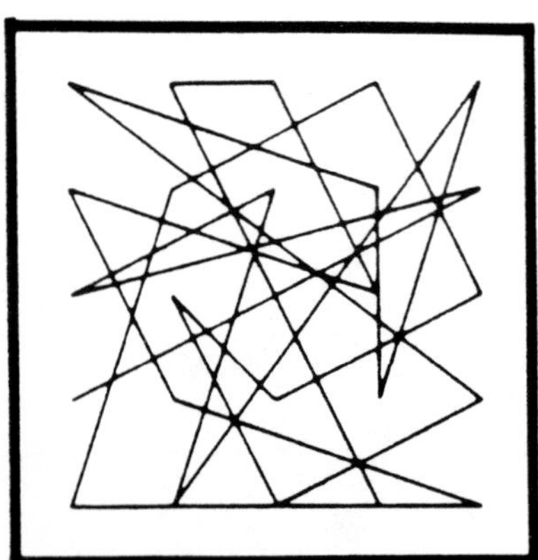

Magic square art can be created by connecting the center points of a square in numerical order.

Magic squares are classified as odd order or even order, with the even order squares further classified as singly even or doubly even. If the order, N, can be expressed as 2m + 1, where m is an integer, we have a magic square of odd order (3, 5, 7, 9, 11, . . .); if N can be written as 2(2m + 1), we have a singly even order (6, 10, 14, . . .); and if N is of the form 4m, the square is doubly even (4, 8, 12, . . .).

Generating magic squares by a microcomputer is relatively new and interesting. Not only are magic squares fun to construct, they also provide excellent programming exercises for the microcomputer user. Once you have an understanding of the methods of constructing various magic squares, you will find them quite simple to produce, with or without a microcomputer.

8.2 HOW TO MAKE MAGIC SQUARES

There are two basic methods of making magic squares. One applies to odd-cell (uneven-cell) squares – 3, 5, 7, 9, etc. – and the other to even-cell squares – 4, 8, 12, 16, etc. To make an odd-cell magic square, we use the *diagonal-arrow method*; to make an even-cell square, we use the *cross-diagonal method*. These methods are described in Sections 8.3 and 8.4.

8.3 ODD-CELL MAGIC SQUARES

The *De la Loubere procedure* is used to generate any magic square of odd order. For the sake of simplicity, a 5 by 5 magic square is generated in the following illustrations. The reader should keep in mind that this method of construction may be used equally well for generating 3 by 3, 7 by 7, 9 by 9, 11 by 11, etc., magic squares.

1. Place the number 1 in the center box of the first row, as shown at the top left of Figure 8-6.
2. Move in an oblique direction, one square to the right and one square above. This movement results in leaving the top of the box. It is necessary to go to the bottom of the column in which you wanted to place the number. Place the number 2 in this location, as shown at the top right of Figure 8-6.
3. Now move diagonally to the right again and put the number 3 in the next box you enter, as shown at the center left of Figure 8-6.
4. If you continue diagonally to the right, you leave the box on the right side. When this occurs you must go to the extreme left of the row in which you wanted to place the number. After crossing over to the left side of the square, put the number 4 into the appropriate box, as shown at the center right of Figure 8-6.
5. Now, again go up diagonally to the right and place the number 5. This completes the first group of five numbers, as shown at the bottom left of Figure 8-6.

a+5b+2c	a	a+4b+c
a+2b	a+3b+c	a+4b+2c
a+2b+c	a+6b+2c	a+b

All 3 by 3 magic squares adhere to the above pattern and have a magic constant of 3a + 9b + 3c (i.e., three times the middle term).

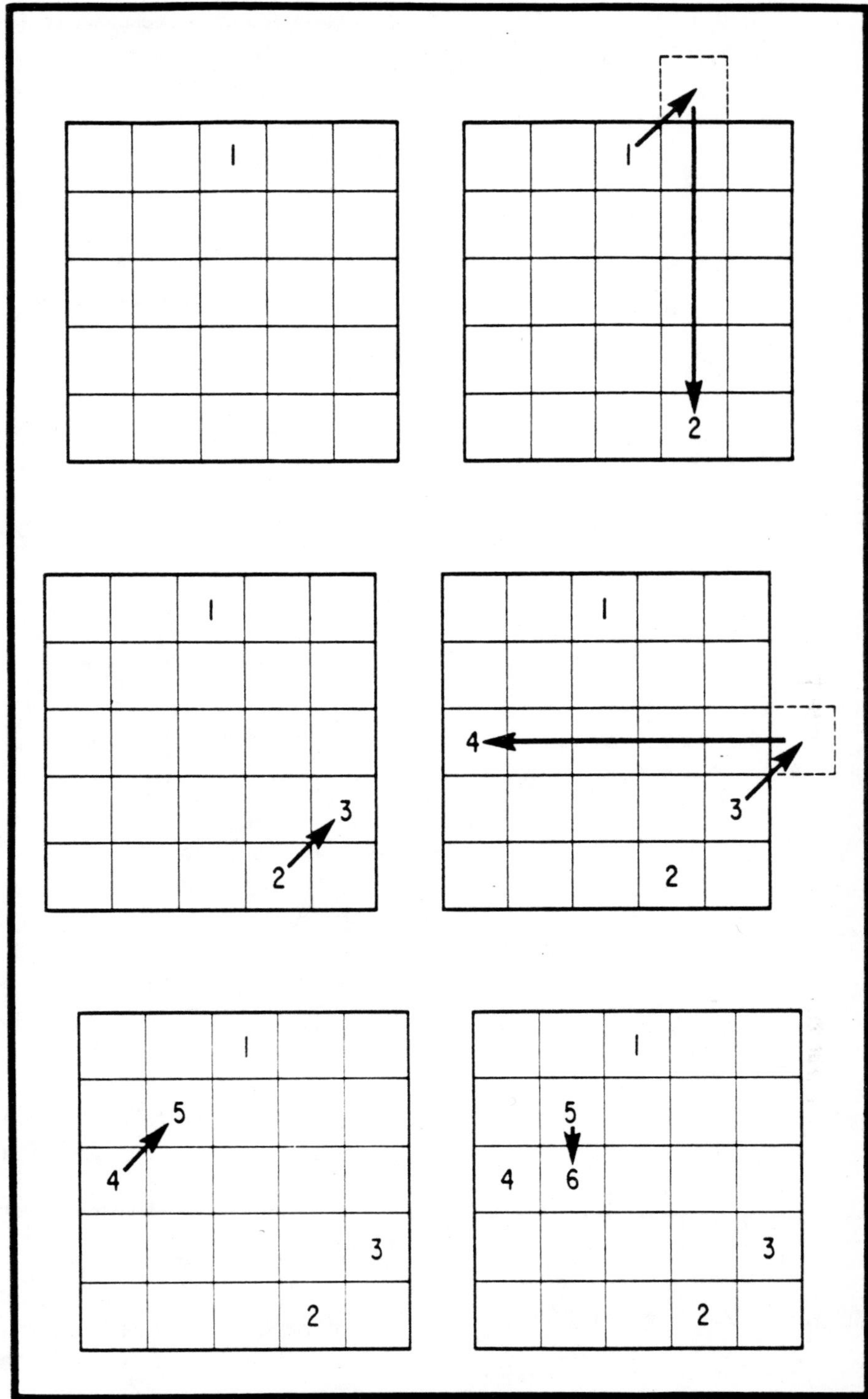

Figure 8-6
A procedure that will generate an odd order magic square (De la Loubere procedure).

6. Since this is a 5 by 5 magic square you must move down one box to generate the next group of five numbers, as shown at the bottom right of Figure 8-6. If this had been a 3 by 3 square or 7 by 7 square, then you would drop down when you reached a group of three or seven numbers, respectively.

7. Move up diagonally to the right, and place a number into each box you enter. If you leave the box at the top, move to the bottom of the column where you wanted to place the number. If

you land outside the box on the right side, move across to the opposite side. After each group of five numbers, go down one box to start the next group of five. When you finish the fifth group of five numbers, the number 25 will occupy the center box of the bottom row, as shown in Figure 8-7.

17	24	1	8	15
23	5	7	14	16
4	6	13	20	22
10	12	19	21	3
11	18	25	2	9

Figure 8-7
An order 5 magic square generated by the De la Loubere procedure.

The sum of each of the five rows, five columns and two main diagonals is 65, and the sum of any two numbers which are diametrically equidistant from the center number is 26, or twice the center number.

Practice constructing this magic square, and then try the next odd order square, order 7 with 49 cells. Exactly the same principle is applied to this order 7 square.

The following program will generate odd-order Magic Squares using the De la Loubere technique.

```
program MagicSquare;
var
  Square        :array[1 .. 25,1 .. 25] of Integer;
  Oddsize       :Integer;(* Size of magic squares *)
  Row,Col       :Integer;(* Row and column indexes *)
  Nextnumber    :Integer;(* Numbers placed in square *)
begin
  (**** Input size of magic square ****)
  write('ENTER SIZE OF SQUARE ');
  readln(Oddsize);
  (**** Store zeros in array Square ****)
  for Row := 1 to Oddsize do
    for Col := 1 to Oddsize do
      Square[Row,Col] := 0;
  (**** Place 1 ****)
  Row := 1;
```

	12	9	
2	7	6	11
5	3	10	8
	4	1	

A **magic cross** is an arrangement of the integers 1 through 12 so that lines across and lines down all add up to the total 26. One example of a magic cross is shown. Can you write a program to generate several other magic crosses?

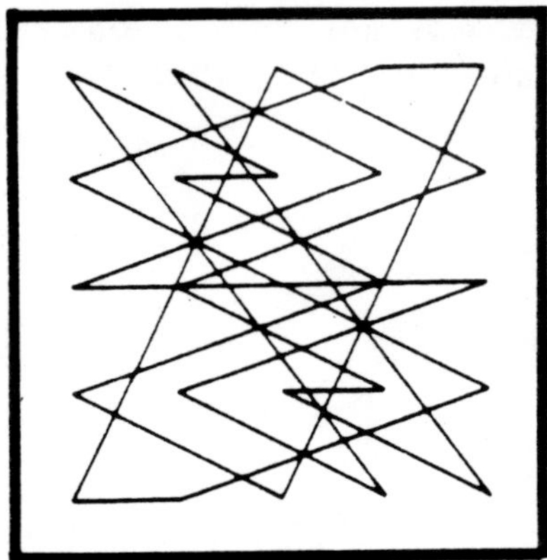

Magic square art

```
Col := Oddsize div 2 + 1;
Square[Row,Col] := 1;
for Nextnumber := 2 to SQR(Oddsize) do
  begin
    Row := Row - 1;
    Col := Col + 1;
    if (Row < 1) and (Col > Oddsize)
      then
        begin
          Row := 2;
          Col := Oddsize;
        end
      else
        if Row < 1
          then Row := Oddsize
          else if Col > Oddsize
            then Col := 1
            else if Square[Row,Col] <> 0
              then
                begin
                  Row := Row + 2;
                  Col := Col - 1;
                end;
   (**** Place next number ****)
   Square[Row,Col] := Nextnumber;
   end;
   (**** Print magic square ****)
   writeln('      5 X 5 MAGIC SQUARE');
   writeln('      ------------------------');
   writeln;
   for Row := 1 to Oddsize do
     begin
       for Col := 1 to Oddsize do
         write(Square[Row,Col]:6);
       writeln;
       writeln;
     end;
end.
```

COMPILE/RUN

```
ENTER SIZE OF SQUARE 5
5 × 5 MAGIC SQUARE
--------------------------
17    24     1     8    15

23     5     7    14    16

 4     6    13    20    22

10    12    19    21     3

11    18    25     2     9
```

NUMBER THEORY TRIVIA

A normal first degree, nth-order magic square is composed of the numbers 1 to n^2, inclusive, arranged in the form of a square so that the n numbers in each row, in each column, and in the two main diagonals, have the same total. This sum, known as the magic constant, equals $n(n^2 + 1)/2$.

A square is said to be **bimagic** if it is magic in both the first and second degrees. In other words, if in addition to meeting the requirements for a normal first degree magic square the second degree square, formed by replacing each of the numbers in the original square by its square, is also magic. The magic constant for a second degree square is $n(n^2 + 1)(2n^2 + 1)/6$.

A square is said to be **trimagic** if it is magic in the first, second, and third degrees. In other words, if in addition to meeting the requirements for a normal bimagic square, the third degree square formed by replacing each of the numbers in the original square by its cube is also magic. The magic constant for a third degree square is $n^3(n^2 + 1)^2/4$.

Can you write a program to generate a 32 by 32 trimagic square? Be aware, however, that generating trimagic squares is quite involved.

A 5 by 5 magic square was generated by the previous program. Figure 8-8 is an order 21 magic square that was generated by this program.

Figure 8-8
An order 21 magic square.

233	256	279	302	325	348	371	394	417	440	1	24	47	70	93	116	139	162	185	208	231
255	278	301	324	347	370	393	416	439	21	23	46	69	92	115	138	161	184	207	230	232
277	300	323	346	369	392	415	438	20	22	45	68	91	114	137	160	183	206	229	252	254
299	322	345	368	391	414	437	19	42	44	67	90	113	136	159	182	205	228	251	253	276
321	344	367	390	413	436	18	41	43	66	89	112	135	158	181	204	227	250	273	275	298
343	366	389	412	435	17	40	63	65	88	111	134	157	180	203	226	249	272	274	297	320
365	388	411	434	16	39	62	64	87	110	133	156	179	202	225	248	271	294	296	319	342
387	410	433	15	38	61	84	86	109	132	155	178	201	224	247	270	293	295	318	341	364
409	432	14	37	60	83	85	108	131	154	177	200	223	246	269	292	315	317	340	363	386
431	13	36	59	82	105	107	130	153	176	199	222	245	268	291	314	316	339	362	385	408
12	35	58	81	104	106	129	152	175	198	221	244	267	290	313	336	338	361	384	407	430
34	57	80	103	126	128	151	174	197	220	243	266	289	312	335	337	360	383	406	429	11
56	79	102	125	127	150	173	196	219	242	265	288	311	334	357	359	382	405	428	10	33
78	101	124	147	149	172	195	218	241	264	287	310	333	356	358	381	404	427	9	32	55
100	123	146	148	171	194	217	240	263	286	309	332	355	378	380	403	426	8	31	54	77
122	145	168	170	193	216	239	262	285	308	331	354	377	379	402	425	7	30	53	76	99
144	167	169	192	215	238	261	284	307	330	353	376	399	401	424	6	29	52	75	98	121
166	189	191	214	237	260	283	306	329	352	375	398	400	423	5	28	51	74	97	120	143
188	190	213	236	259	282	305	328	351	374	397	420	422	4	27	50	73	96	119	142	165
210	212	235	258	281	304	327	350	373	396	419	421	3	26	49	72	95	118	141	164	187
211	234	257	280	303	326	349	372	395	418	441	2	25	48	71	94	117	140	163	186	209

8.4 EVEN-CELL MAGIC SQUARES

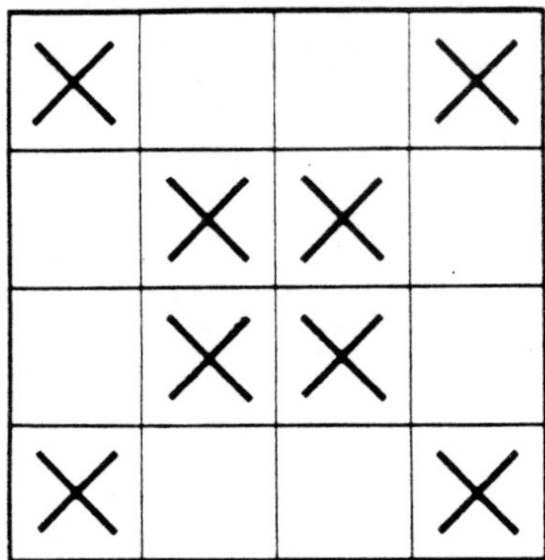

Figure 8-9
First step in the generation of an even-cell magic square.

The following steps show the generation of an even-cell magic square of four cells:

1. In a blank 4 by 4 square, fill the main diagonal squares with X's, as in Figure 8-9.

2. Start with the upper left square and move toward the right, obeying the following rules: (a) If the cell is occupied by an X, skip the cell; and (b) if the cell is not occupied by an X, insert a number. Start with the number 1 and increment your count by 1 each time a move is made. On reaching the end of a row, repeat the process in the next row.

3. The first eight numbers would be placed in the square as shown in Figure 8-10.

4. Now fill in the cells containing an X. Start in the same cell as in step 2 (upper left square) and obey the following rules: (a) If the cell is occupied by an X, insert a number; and (b) if the cell is occupied by a number, skip the cell. Start with the number 16 and decrease the count by 1 each time a move is made. When the end of a row is reached, repeat the same process in the next row.

5. The last eight numbers would be placed in the square as shown in Figure 8-11.

6. The completed magic square would appear as shown in Figure 8-12.

The following program uses this procedure to generate a 4 by 4 magic square.

```
program EvenOrderMagicSquare;
var
  N       :Integer;(* Size of square to be generated *)
  I,J     :Integer;(* Row and Column subscripts *)
  K       :Integer;(* Integers being placed in array *)
  M       :array[1..4,1..4] of Integer;(* Magic square array *)
begin  N := 4;
   (**** Store zeros in array M ****)
   for I := 1 to N do
     for J := 1 to N do
       M[I,J] := 0;
   (**** Store 999 in each cell of diagonal 1 ****)
   for I := 1 to N do
     begin
       J := I;
       M[I,J] := 999;
```

3	61	19	37
43	31	5	41
7	11	73	29
67	17	23	13

An order 4 magic square composed of prime numbers.

	2	3	
5			8
9			12
	14	15	

CELL	X's	CONTENTS OF SQUARE
1	X	NOT CHANGED
2		2 → CELL 2
3		3 → CELL 3
4	X	NOT CHANGED
5		5 → CELL 5
6	X	NOT CHANGED
7	X	NOT CHANGED
8		8 → CELL 8
9		9 → CELL 9
10	X	NOT CHANGED
11	X	NOT CHANGED
12		12 → CELL 12
13	X	NOT CHANGED
14		14 → CELL 14
15		15 → CELL 15
16	X	NOT CHANGED

Figure 8-10
First eight numbers are placed in the square as shown.

16			13
	11	10	
	7	6	
4			1

CELL	X's	CONTENTS OF SQUARE
1	X	16 → CELL 1
2		NOT CHANGED
3		NOT CHANGED
4	X	13 → CELL 4
5		NOT CHANGED
6	X	11 → CELL 6
7	X	10 → CELL 7
8		NOT CHANGED
9		NOT CHANGED
10	X	7 → CELL 10
11	X	6 → CELL 11
12		NOT CHANGED
13	X	4 → CELL 13
14		NOT CHANGED
15		NOT CHANGED
16	X	1 → CELL 16

Figure 8-11
Second eight numbers are placed in the square as shown.

16	2	3	13
5	11	10	8
9	7	6	12
4	14	15	1

Figure 8-12
An order 4, even-cell magic square.

```
    end;
  (**** Store 999 in each cell of diagonal 2 ****)
  for I := 1 to N do
    begin
      J := N - I + 1;
      M[I,J] := 999;
    end;
  (**** First pass through the array ****)
  K := 1;
  for I := 1 to N do
    for J := 1 to N do
      begin
        IF M[I,J] = 0 then M[I,J] := K;
        K := K + 1;
      end;
  (**** Second pass through the array ****)
  K := N * N;
  for I := 1 to N do
    for J := 1 to N do
      begin
        if M[I,J] = 999 then M[I,J] := K;
        K := K - 1;
      end;
  (**** Print magic square ****)
  writeln('   4 BY 4 MAGIC SQUARE');
  writeln;
  for I := 1 to N do
    begin
      for J := 1 to N do
        write(M[I,J]:6);
        writeln;
        writeln;
      end;
end.
```

COMPILE/RUN

```
4 BY 4 MAGIC SQUARE
16    2    3   13

 5   11   10    8

 9    7    6   12

 4   14   15    1
```

277	197	631	431
661	401	307	167
137	337	491	571
461	601	107	367

Shown above is an order 4 pandiagonal magic square composed of 3-digit prime numbers. The magic cosntant of this square is 1536.

Adding 30 to each of the primes in the above square produces another order 4 pandragonal magic square, composed of 3-digit primes. This new magic square has a magic constant of 1656.

307	227	661	461
691	431	337	197
167	367	521	601
491	631	137	397

Write a program that will add an integer less than 1100 to each prime in the above square. The resulting square should be an order 4 pandiagonal magic square with a magic constant of 6024.

The next magic square of doubly even order is of order 8. It can be constructed in a manner similar to the method used previously for the order 4 magic square. First, consider that the order 8 magic square is subdivided into four squares of order 4, and then visualize diagonals filled with X's drawn in each of these order four squares. Figure 8-13 illustrates how the diagonals are marked by placing X's on all diagonals of the order 4 squares. The procedure applied to the generation of order 4 squares also applies to the generation of order 8 squares, but the number range includes the numbers 1 through 64. You should construct an order 8 magic square using the procedure shown in Figure 8-13.

Figure 8-13
Procedure for generating an order 8 magic square.

X			X	X			X
	X	X			X	X	
	X	X			X	X	
X			X	X			X
X			X	X			X
	X	X			X	X	
	X	X			X	X	
X			X	X			X

A blank 8 x 8 square with X's on the diagonals of each 4 x 4 square.

	2	3			6	7	
9			12	13			16
17			20	21			24
	26	27			30	31	
	34	35			38	39	
41			44	45			48
49			52	53			56
	58	59			62	63	

The blank cells are filled in with numbers. Start in the upper left cell with 1 and move to the right increasing the count with each move.

64			61	60			57
	55	54			51	50	
	47	46			43	42	
40			37	36			33
32			29	28			25
	23	22			19	18	
	15	14			11	10	
8			5	4			1

The cells containing X's are filled in with numbers. Start in the upper left cell with 64 and move toward the right decreasing the count with each move.

64	2	3	61	60	6	7	57
9	55	54	12	13	51	50	16
17	47	46	20	21	43	42	24
40	26	27	37	36	30	31	33
32	34	35	29	28	38	39	25
41	23	22	44	45	19	18	48
49	15	14	52	53	11	10	56
8	58	59	5	4	62	63	1

Generated 8 x 8 Magic Square

n−8	n+7	n+6	n−5
n+4	n−3	n−2	n+1
n−1	n+2	n+3	n−4
n+5	n−6	n−7	n+8

Shown above is a general form for an order 4 magic square with a magic constant equal to 4n.

8.5 WHAT NUMBERS WILL MAGIC SQUARES ADD UP TO?

We have seen that a magic square of order 3 (3 by 3 cells) has all rows and columns adding up to 15 when numbers from 1 to 9 are used. We have seen that a magic square of order 4 (4 by 4 cells) has rows and columns adding up to 34 when all numbers from 1 to 16 are used. What about squares of order 5, 6, 7, or larger? What will the rows and columns of these add up to? The sum of the rows, columns, and main diagonals, called the *magic constant*, is determined by the formula,

$$\text{Magic constant} = n(n^2 + 1)/2$$

where n is the order of the square. For example, in the magic square of order 3(9 cells), the magic constant is determined as follows:

$$\text{Magic constant} = 3(3^2 + 1)/2 = 3(9 + 1)/2 = 30/2 = 15$$

In an order 5 square, n = 5:

$$5(5^2 + 1)/2 = 5(25 + 1)/2 = 5(26)/2 = 130/2 = 65$$

That is, all rows, columns and main diagonals add up to 65.

In a square with 25 rows and 25 columns, n = 25:

$$25(25^2 + 1)/2 = 25(625 + 1)/2 = 25(626)/2 = 15650/2 = 7825$$

8.6 MAGIC SQUARES STARTING WITH NUMBERS OTHER THAN ONE

The magic squares previously discussed all started with the number 1. However, magic squares may be started with any number. The 3 by 3 square in Figure 8-14 starts with 4.

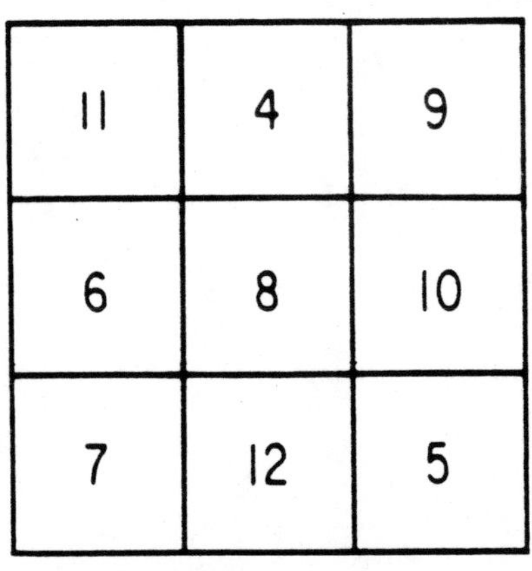

11	4	9
6	8	10
7	12	5

Figure 8-14
An order 3 magic square that starts with the number 4.

The magic constant of this square is 24 and is computed by the formula

$$\text{magic constant} = \frac{N^3 + N}{2} + N(P - 1) = \frac{3^3 + 3}{2} + 3(4 - 1) = 24$$

where N is the order of the square and P is the starting number.

Figure 8-15 shows a 15 by 15 magic square that starts with 7. The magic constant of this square is 1785.

The following program will generate an odd order magic square starting with any number.

Figure 8-15
An order 15 magic square that starts with the number 7.

128	145	162	179	196	213	230	7	24	41	58	75	92	109	126
144	161	178	195	212	229	21	23	40	57	74	91	108	125	127
160	177	194	211	228	20	22	39	56	73	90	107	124	141	143
176	193	210	227	19	36	38	55	72	89	106	123	140	142	159
192	209	226	18	35	37	54	71	88	105	122	139	156	158	175
208	225	17	34	51	53	70	87	104	121	138	155	157	174	191
224	16	33	50	52	69	86	103	120	137	154	171	173	190	207
15	32	49	66	68	85	102	119	136	153	170	172	189	206	223
31	48	65	67	84	101	118	135	152	169	186	188	205	222	14
47	64	81	83	100	117	134	151	168	185	187	204	221	13	30
63	80	82	99	116	133	150	167	184	201	203	220	12	29	46
79	96	98	115	132	149	166	183	200	202	219	11	28	45	82
95	97	114	131	148	165	182	199	216	218	10	27	44	61	78
111	113	130	147	164	181	198	215	217	9	26	43	60	77	94
112	129	146	163	180	197	214	231	8	25	42	59	76	93	110

```
program AnyNumberMagicSquare;
var
  Square       :array[1 .. 25,1 .. 25] of Integer;
  Oddsize      :Integer;(* Size of magic square *)
  Start        :Integer;(* Starting number of square *)
  Row,Col      :Integer;(* Row and column indexes *)
  N            :Integer;(* Looping counter *)
  Nextnumber   :Integer;(* Numbers placed in square *)
begin
  (**** Input size of magic square ****)
  write(' ENTER SIZE OF SQUARE: ');
  readln(Oddsize);
  (**** Input starting number of square ****)
  write(' ENTER STARTING NUMBER: ');
  readln(Start);
  Nextnumber := Start;
  (**** Store zeros in array Square ****)
  for Row := 1 to Oddsize do
    for Col := 1 to Oddsize do
      Square[Row,Col] := 0;
  (**** Place first number in square ****)
  Row := 1;
  Col := Oddsize div 2 + 1;
```

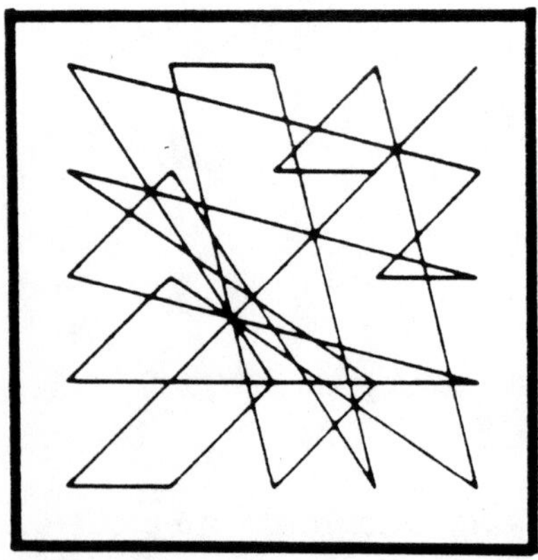

Magic square art

```
Square[Row,Col] := Start;
for N := 2 to SQR(Oddsize) do
  begin
    Row := Row - 1;
    Col := Col + 1;
    if(Row < 1) and (Col > Oddsize)
      then
        begin
          Row := 2;
          Col := Oddsize;
        end
      else
        if Row < 1
          then Row := Oddsize
          else if Col > Oddsize
            then Col := 1
            else if Square[Row,Col] < > 0
              then
                begin
                  Row := Row + 2;
                  Col := Col - 1;
                end.
  (**** Place next number ****)
  Nextnumber := Nextnumber + 1;
  Square[Row,Col] := Nextnumber;
end;
(**** Print magic square ****)
writeln;
write('    ',Oddsize,' BY ',Oddsize,' MAGIC SQUARE');
writeln(' STARTING WITH ',Start);
writeln;
for Row := 1 to Oddsize do
  begin
    for Col := 1 to Oddsize do
      write(Square[Row,Col]:6);
    writeln;
    writeln;
  end;
end.
```

COMPILE/RUN

Note: Program output is shown on page 170.

101	29	83
53	71	89
59	113	41

Shown above is a magic square consisting of non-consecutive prime numbers. The magic constant is 213. Can you find an order 3 magic square that is composed of consecutive prime numbers? Harry Nelson of California used a Cray supercomputer to determine the following magic square composed of consecutive primes.

1480028201	1480028129	1480028183
1480028153	1480028171	1480028189
1480028159	1480028213	1480028141

5

6 2

8 4

1 3 7 9

Shown here is an Order 4 *magic triangle.* The numbers 1 through 9 are arranged in a triangular array in such a way that the three sides add to the same number. Can you write a program to generate other magic triangles?

```
ENTER SIZE OF SQUARE: 7
ENTER STARTING NUMBER: 612

7 BY 7 MAGIC SQUARE STARTING WITH 612

641   650   659   612   621   630   639
649   658   618   620   629   638   640
657   617   619   628   637   646   648
616   625   627   636   645   647   656
624   626   635   644   653   655   615
632   634   643   652   654   614   623
633   642   651   660   613   622   631
```

Input to this program is the order of the square and the starting number. In the example shown, 7 is the order and 612 is the starting number.

After receiving the input information, the program causes the following heading to be typed:

7 BY 7 MAGIC SQUARE STARTING WITH 612

8.7 MULTIPLICATION MAGIC SQUARE

A 3 by 3 multiplication magic square is shown in Figure 8-16. The magic constant, obtained by multiplying together the three numbers in any column, row, or diagonal, is 216.

A method based on the De la Loubere odd order constructing method may be used to generate multiplication magic squares of odd order. The construction of a 5 by 5 multiplication square will be used to illustrate the method.

1. Place the number 1 in the center cell of the first row in a blank 5 by 5 square, as shown at the top left of Figure 8-17.
2. Move in an oblique direction, one square above and to the right. This movement results in leaving the top of the box. It is necessary to place the next number in a cell at the bottom of the column in which you attempted to place the number. The number to place at this location is twice the last number, or 2, as shown at the top right of Figure 8-17.

2. Move in an oblique direction, one square above and to the right. This movement results in leaving the top of the box. It is necessary to place the next number in a cell at the bottom of the column in which you attempted to place the number. The number to place at this location is twice the last number, or 2, as shown at the top right of Figure 8-17.

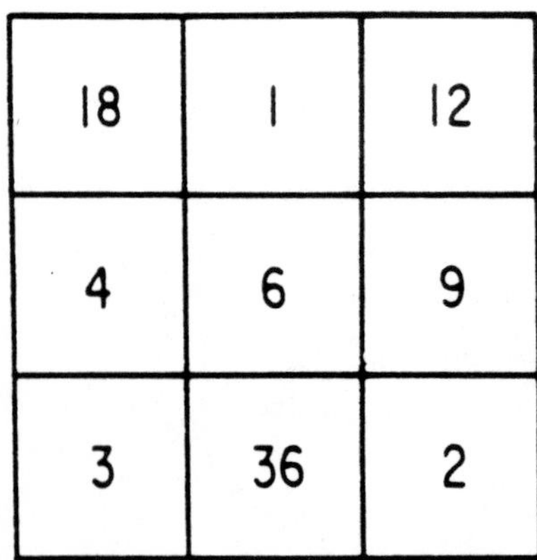

18	1	12
4	6	9
3	36	2

Figure 8-16
An order 3 multiplication magic square.

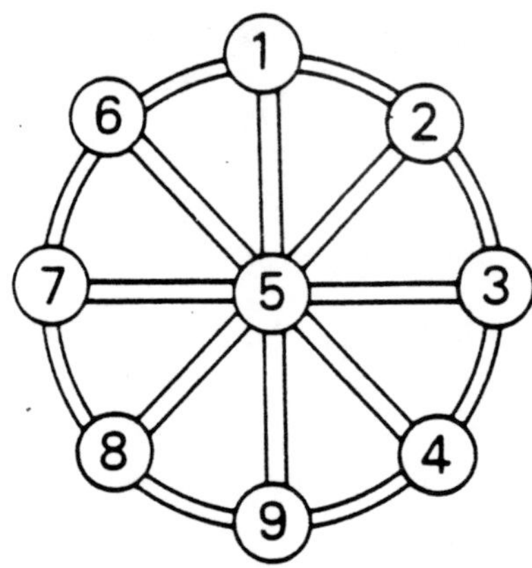

A **magic wheel** is an arrangement of the numbers 1 through 9 around the nodes so that lines across the wheel — from outer node to the center node and on to the opposite outer node — all add to the total 15. An example of a magic wheel is shown. Can you write a program to generate several other magic wheels?

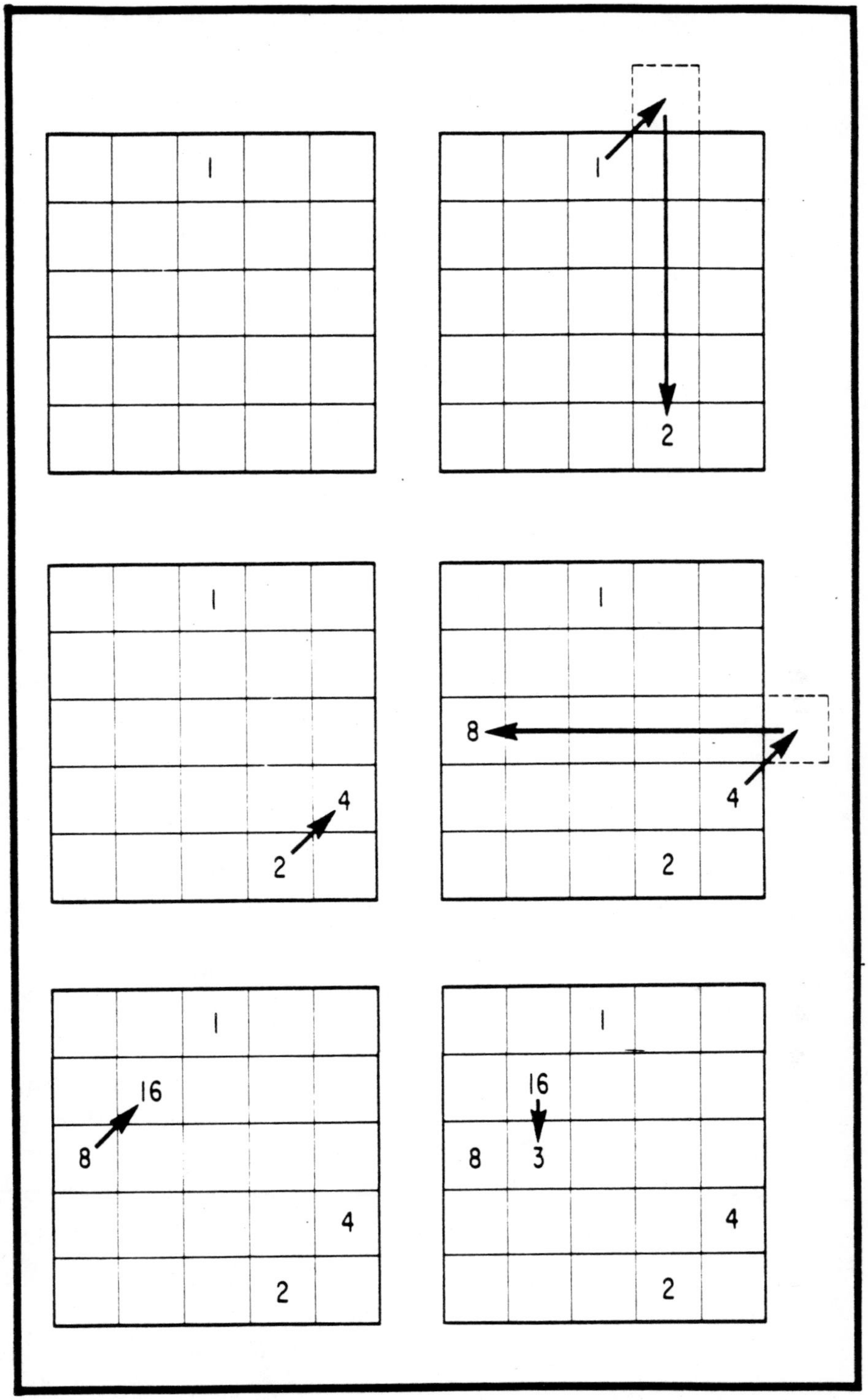

Figure 8-17
Starting procedure for generating an order 5 multiplication magic square.

3. Now move diagonally to the right again, and put a number twice as large as the last, or 4, in the next cell location, as shown at the center left of Figure 8-17.

4. If you continue diagonally to the right, you will leave the cell on the right side. When this occurs, you must go to the extreme left of the row in which you attempted to place the next number. After crossing over to the left side of the square,

NUMBER THEORY TRIVIA

There are 880 different (excluding rotations and reflections) magic squares of Order 4. There are 275,305,224 different magic squares of Order 5.

place a number twice that of the last, or 8, into the appropriate cell, as shown at the center right of Figure 8-17.

5. Now, again, go up diagonally to the right and place the next number. This number, determined by doubling the previous number, is 16, as shown at the bottom left of Figure 8-17.

   This completes the first group of five numbers of a 5 by 5 square. The next group of five numbers starts with 3, the next group 9, the next group 27, and the last group 81. The reader should note that the starting numbers are all powers of 3:

$$3^0 = 1$$
$$3^1 = 3$$
$$3^2 = 9$$
$$3^3 = 27$$
$$3^4 = 81$$

6. Since this is a 5 by 5 square, you must move down one cell to place the next group of five numbers. In the case of a 3 by 3 square, you would move down when you reached a group of 3. The number to be placed in this cell is the starting number of the second group of 5 numbers, or 3, as shown at the bottom right of Figure 8-17.

7. Move up diagonally to the right and place a number into each cell you enter, always doubling the previous number. When you leave the top of the box, move to the bottom of the column where you attempted to place the number. When you move outside the box on the right side, move across to the opposite side. When you finish the second group of five numbers, the square should appear as shown at the top left of Figure 8-18.

8. The square at the top right of Figure 8-18 would result after the third group of five numbers have been placed. The starting number is 9.

9. The fourth group of five numbers would be placed in the manner shown at the bottom left of Figure 8-18. The starting number is 27.

10. When the twenty-fifth number is placed in the cell opposite the starting cell, the final square will appear as shown at the bottom right of Figure 8-18.

Perhaps the reader would like to write a program to generate a multiplication magic square.

6 14 15 3 13
8 1 12 10
7 11 2
4 9
5

Shown here is an **absolute difference triangle**. It can be constructed by placing below each consecutive pair the absolute value of their difference, and continuing the process until a single number is obtained. Can you write a program to generate several absolute difference triangles?

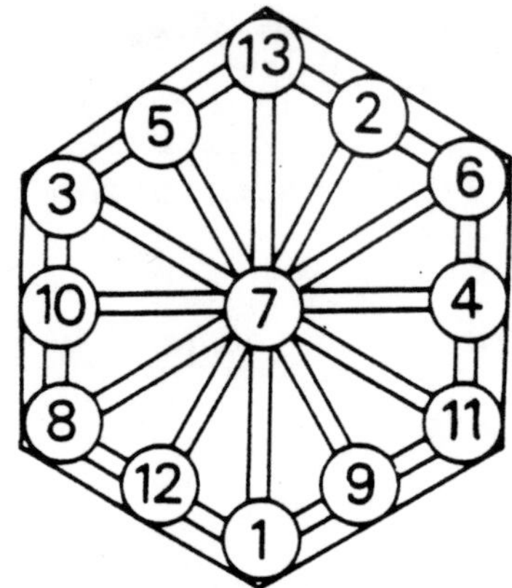

A **magic hexagon** is an arrangement of the numbers 1 through 13 around the nodes so that lines — from outer node to the center node and on to the opposite outer node, and also along each side — total 21. One example of a magic hexagon is shown. Can you write a program to generate several other magic hexagons?

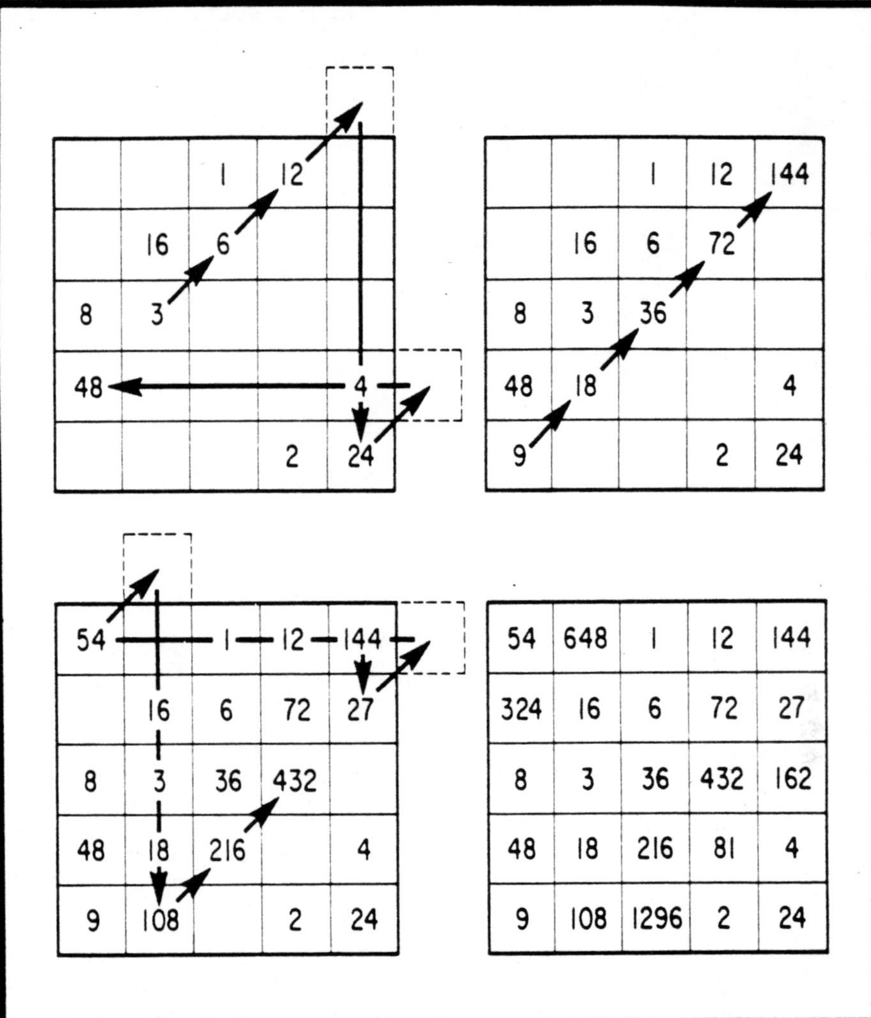

Figure 8-18
Final steps in the generaton of an order 5 multiplication magic square.

8.8 GEOMETRIC MAGIC SQUARE

A geometric magic square is an array of numbers where the product of the numbers in every column, row, and main diagonal is the same and where each number of the square is represented by a base

Figure 8-19
Order 3 geometric magic square.

2^8	2^1	2^6
2^3	2^5	2^7
2^4	2^9	2^2

GEOMETRIC MAGIC SQUARE USING BASE AND EXPONENT VALUES.

256	2	64
8	32	128
16	512	4

GEOMETRIC MAGIC SQUARE USING INTEGER VALUES.

value and an exponent. The base value remains the same in all the positions of the square, and the exponent values are the numbers in an ordinary odd order magic square. For example, an order 3 geometric magic square with a base of 2 would appear as shown in Figure 8-19.

A program for generating a geometric magic square is shown below.

```
program GeometricMagicSquare;
var
  Square       :array[1 .. 25,1 .. 25] of Integer;
  Oddsize      :Integer;(* Size of square to be generated *)
  B            :Real;(* Base of square *)
  Row,Col      :Integer;(* Indexes to square *)
  N            :Integer;(* Looping counter *)
  P            :Real;(* Power of base *)
function power(No,Ex:Real):Real;
  begin
    Power := EXP(LN(B) * P);
  end;
begin
  (**** Input size of square to be generated ****)
  write('ENTER SIZE OF SQUARE TO BE GENERATED ');
  readln(Oddsize);
  (**** Input base of square ****)
  write('ENTER BASE OF SQUARE ');
  readln(B);
  (**** Store zeros in array Square ****)
  for Row := 1 to Oddsize do
    for Col := 1 to Oddsize do
      Square[Row,Col] := 0;
  P := 1.0;
  (**** Place first number in square ****)
  Row := 1;
  Col := Oddsize div 2 + 1;
  Square[Row,Col] := TRUNC(B);
  for N := 2 to TRUNC(Sqr(Oddsize)) do
    begin
      P := P + 1;
      Row := Row - 1;
      Col := Col + 1;
      if(Row < 1) and (Col > Oddsize)
        then
          begin
            Row := 2;
            Col := Oddsize;
          end
        else
          if Row < 1
            then Row := Oddsize
            else if Col > Oddsize
```

```
                    then Col := 1
                    else if Square[Row,Col] <> 0
                      then
                        begin
                          Row := Row + 2;
                          Col := Col - 1;
                        end;
        (**** Place next number ****)
        Square[Row,Col] := TRUNC(Power(B,P));
      end;
      (**** Print geometric magic square ****)
      writeln;
      writeln('3 BY 3 GEOMETRIC MAGIC SQUARE');
      writeln;
      for Row := 1 to Oddsize do
        begin
          for Col := 1 to Oddsize do
            write(Square[Row,Col]:6);
          writeln;
          writeln;
        end;
end.
```

```
COMPILE/RUN

ENTER SIZE OF SQUARE TO BE GENERATED 3
ENTER BASE OF SQUARE 2

3 BY 3 GEOMETRIC MAGIC SQUARE

256     2    64

  8    32   128

 16   512     4
```

Figure 8-20 illustrates two other order 3 geometric magic squares. One of the squares uses a base of 4 and the other a base of 5.

65536	4	4096
64	1024	16384
256	262144	16

390625	5	15625
125	3125	78125
625	1953125	25

Figure 8-20
Geometric magic squares using bases 4 and 5.

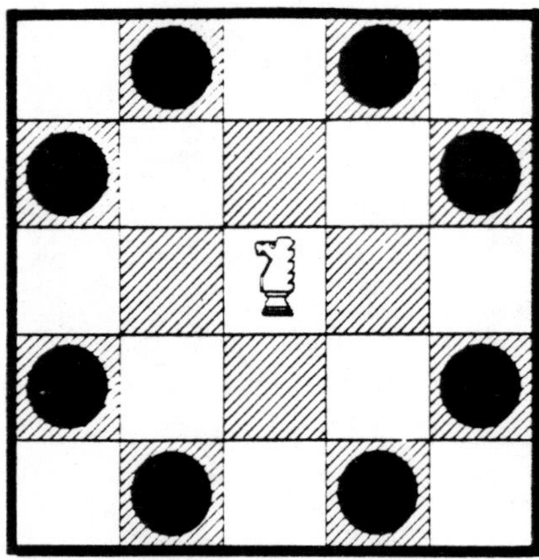

Moves permitted a knight

8.9 OTHER INTERESTING MAGIC SQUARES

A square, whose magic constant is 2056, is shown in Figure 8-21. The unusual nature of this square lies in its method of construction. If the numbers are followed consecutively, it is found that the "moves" from one to the next are the moves of a knight in the game of chess.

Figure 8-21
An order 16 magic square constructed with moves of the knight.

184	217	170	75	188	219	172	77	228	37	86	21	230	39	88	25
169	74	185	218	171	76	189	220	85	20	229	38	87	24	231	40
216	183	68	167	222	187	78	173	36	227	22	83	42	237	26	89
73	168	215	186	67	174	221	190	19	84	35	238	23	90	41	232
182	213	166	69	178	223	176	79	226	33	82	31	236	43	92	27
165	72	179	214	175	66	191	224	81	18	239	34	91	30	233	44
212	181	70	163	210	177	80	161	48	225	32	95	46	235	28	93
71	164	211	180	65	162	209	192	17	96	47	240	29	94	45	234
202	13	126	61	208	15	128	49	160	241	130	97	148	243	132	103
125	60	203	14	127	64	193	16	129	112	145	242	131	102	149	244
12	201	62	123	2	207	50	113	256	159	98	143	246	147	104	133
59	124	11	204	63	114	1	194	111	144	255	146	101	134	245	150
200	9	122	55	206	3	116	51	158	253	142	99	154	247	136	105
121	58	205	10	115	54	195	4	141	110	155	254	135	100	151	248
8	199	56	119	6	197	52	117	252	157	108	139	250	153	106	137
57	120	7	198	53	118	5	196	109	140	251	156	107	138	249	152

83	29	101
89	71	53
41	113	59

Figure 8-22
An order 3 magic square composed of prime numbers.

The magic square of Figure 8-22 is composed entirely of prime numbers.

A square that is magic for both addition and multiplication is shown in Figure 8-23. The magic constant for addition is 1200, while the multiplication constant is 1,619,541,385,529,760,000.

46	55	44	19	58	9	22	7
43	18	47	56	21	6	59	10
54	45	20	41	12	57	8	23
17	42	53	48	5	24	11	60
52	3	32	13	40	61	34	25
31	16	49	4	33	28	37	62
2	51	14	29	64	39	26	35
15	30	1	50	27	36	63	38

An order 8 magic square constructed with moves of the knight. The magic constant of this square is 260.

17	171	126	54	230	100	93	264	145
124	66	290	85	57	168	162	23	225
216	115	75	279	198	29	170	76	42
261	186	33	210	68	38	200	135	69
50	270	92	87	248	165	21	153	114
105	51	152	150	27	207	116	62	330
138	25	243	132	58	310	95	63	136
190	84	34	184	125	81	297	174	31
99	232	155	19	189	102	46	250	108

Figure 8-23
An order 9 square that is completely magic for both addition and subtraction.

8.10 HETEROSQUARES

A heterosquare is an N by N array of integers from 1 to N^2 such that all the rows, columns and diagonals have different sums.

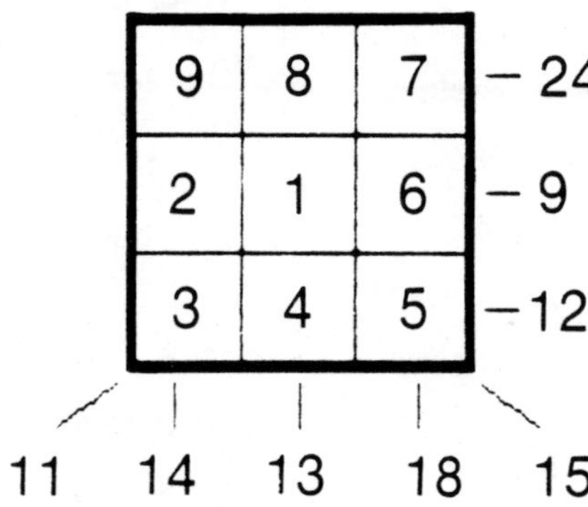

49	48	47	46	45	44	43
26	25	24	23	22	21	42
27	10	9	8	7	20	41
28	11	2	1	6	19	40
29	12	3	4	5	18	39
30	13	14	15	16	17	38
31	32	33	34	35	36	37

Figure 8-24
An order 7 heterosquare.

A spiral technique can be used to produce heterosquares. One starts from the center and places consecutive integers in a spiral, as shown in figure 8-24.

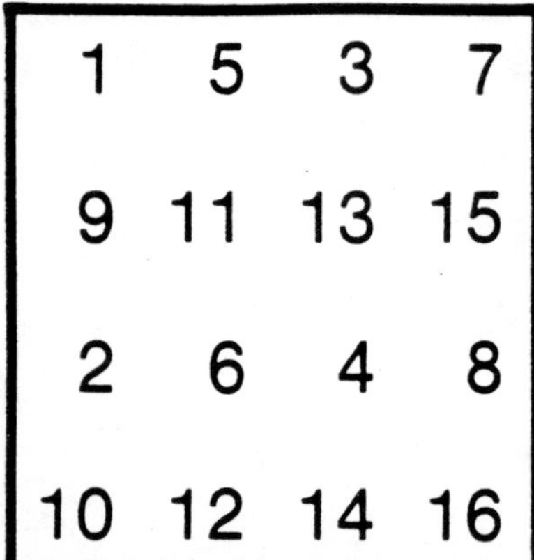

1	5	3	7
9	11	13	15
2	6	4	8
10	12	14	16

Figure 8-25
A talisman square in which the differences are greater than 1.

5	15	9	12
10	1	6	3
13	16	11	14
2	8	4	7

Figure 8-26
A talisman square in which the differences are greater than 2.

8.11 TALISMAN SQUARES

A talisman square is an N by N array of integers from 1 to N^2 so that the difference between any one integer and its neighbor is greater than some given constant. A neighboring integer is one immediately adjacent to a given number either vertically, horizontally, or diagonally. Figure 8-25 shows a talisman square in which the difference between any integer and its neighbor is greater than one. The integer 10, for example, has only three neighbors (2, 6, 12); 14 has five neighbors (12, 6, 4, 8, 16); and 6 has eight neighbors (10, 2, 9, 11, 13, 4, 14, 12). Figure 8-26 shows a talisman square in which the differences are greater than 2.

The study of these squares is so new in fact, that no rules for construction are known, nor are there any mathematical theories that can help in deciding the maximum possible difference between integers and their neighbors. Among the many questions still requiring answers are: What are the properties of talisman squares? Are there any systematic methods by which talisman squares can be constructed? What is the maximum possible difference between integers and their neighbors for a given Nth-order talisman square? Can talisman squares find practical application?

REVIEW EXERCISES

1. Show that the sum of the rows of the lo-shu square is equal to the sum of the columns.
2. Show that the sum of the numbers of the second and fourth columns of the Melencolia magic square is equal to the sum of the numbers of the first and third rows.
3. What is the magic constant of the magic square of order (a) 6? (b) 7? (c) 15? (d) 21? Give an example of (e) a singly even order magic square; (f) a doubly even order magic square.
4. Hand construct a magic square of order 3.
5. The De la Loubere procedure is used to generate magic squares of an ______________ order.
6. Hand construct a magic square of order 5.
7. Hand construct a magic square of order 7.
8. Hand construct a magic square of order 9.
9. Hand construct a magic square of order 11.
10. Hand construct a magic square of order 13.
11. Hand construct a magic square of order 15.
12. Hand construct a magic square of order 17.
13. Use the program shown in Section 8.3 to generate an order 7 magic square.

3	5	14	12
10	16	7	1
8	2	9	15
13	11	4	6

Shown here is a 4 by 4 square that is both magic and talismanic with talisman constant 2. Can you write a computer program to produce 23 additional 4 by 4 squares that use the numbers 1, 2, 3, . . ., 16?

14. Use the program shown in Section 8.3 to generate an order 9 magic square.

15. Use the program shown in Section 8.3 to generate an order 11 magic square.

16. Use the program shown in Section 8.3 to generate an order 13 magic square.

17. Hand construct a magic square of order 8.

18. Hand construct a magic square of order 12. (Think of it as being subdivided into nine magic squares of order 4. Then proceed as we did in the construction of an order 8 magic square).

19. What is a magic constant?

20. What is the magic constant of an order 9 magic square?

21. What is the magic constant of a 21 by 21 magic square?

22. The Emperor Charlemagne (742-814) ordered a five-sided fort to be built at an important point in his kindgom. As good-luck charms, he had magic squares placed on all five sides of the fort. He had one restriction for these magic squares: all the numbers in them must be prime. These magic squares are listed below. Find the magic constant for each square. You should get the same magic number for each of the five squares. This number gives the year that the fort was built.

479	71	257
47	269	491
281	467	59

389	191	227
107	269	431
311	347	149

389	227	191
71	269	467
347	311	149

401	227	179
47	269	491
359	311	137

401	257	149
17	269	521
389	281	137

23. What is the magic constant of an order 279 magic square?

24. Write a program to generate an order 8 magic square.

25. Use the program shown in Section 8.6 and generate an order 5 magic square starting with 18.

26. Use the program shown in Section 8.6 and generate an order 7 magic square that starts with 23.

27. Use the program shown in Section 8.6 to produce an order 11 magic square that starts with 15.

28. Write a program to produce an order 5 multiplication magic square.

29. How does the multiplication magic square differ from a regular magic square?

30. Use the program shown in Section 8.8 to generate an order 3 geometric magic square with base 4.

31. In what way does the geometric magic square differ from the multiplication magic square?

6	1	8
7	5	3
2	9	4

Lo-shu magic square.

32. Magic squares can be transformed into other magic squares by adding, multiplying, or dividing all the numbers by any constant value. Transform the lo-shu magic square by first adding 2 to each cell and then dividing by 8. Show the results in the form of simplified fractions. Then verify that the new square is, indeed, another magic square.

33. Connecting the center points of the 16 squares in Durer's magic square in numerical order forms an interesting symmetric design. Try it. Shade portions of the diagram to produce an original piece of magic square art.

34. Select a magic square and design some original magic square art.

35. Write a program to determine if the following square is a magic multiplication square.

15	16	22	3	9
8	14	20	21	2
1	7	13	19	25
24	5	6	12	18
12	23	4	10	11

36. Write a program that will generate a heterosquare.

37. What is a talisman square?

7	12	1	14
2	13	8	11
16	3	10	5
9	6	15	4

The oldest known 4 by 4 magic square is from an 11th century carving at Khajuraho in India.

9 NUMBER SYSTEMS

PREVIEW

Until the advent of computers, the decimal system reigned supreme in all fields of numerical calculations. The interest expressed in other systems was mainly historical and cultural. Since computers operate on binary numbers, and programs sometimes contain octal or hexadecimal numbers, it is appropriate for students of number theory to understand these numeration systems.

After you finish this chapter, you should be able to:

1. Identify the number bases used with computers.
2. Represent numbers in decimal, binary, hexadecimal and octal notations.
3. Convert numbers from one number base to equivalent numbers in another number base.
4. See how a microcomputer can be used to convert numbers from one base to another base.

9.1 INTRODUCTION

Most modern computers use the binary digit as a basic unit of information. The reason for this is that the two digits of the binary number system are easier to represent and use electronically than are the ten Arabic numerals of the decimal system. However, since people normally use the decimal number system, it is necessary to provide a method for the computer user to communicate with the computer. This usually results in the computer using an internal code of binary notation, whereas input and output information to the computer is handled in either decimal, octal, or hexadecimal notation. Of course, this involves a translation of the input and output information.

In most cases, it will be sufficient for one to think and use decimal values. In some cases, however, such as shifting, scaling numbers, logical operations, reading memory dumps, program debugging, etc., a knowledge of binary, octal, and hexadecimal representation is necessary.

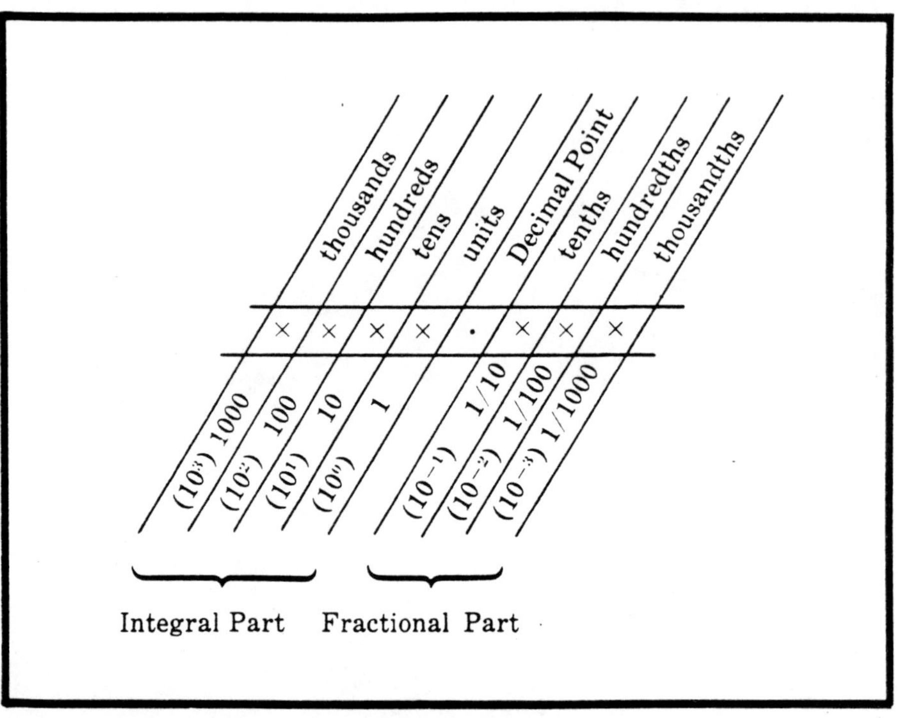

The positional weights of a decimal number (powers of 10).

Every number system has three concepts in common: (1) a base, or radix, (2) digit value, and (3) positional notation. The base is the number of different digits used in that system. Each digit of a specified number system has a distinct value. Each number position carries a specific weight depending on the base of the system. For example, the decimal number 3,417 can be described as follows:

3 4 1 7

$$7 \times 10^0 = 7 \times 1 = 7$$
$$1 \times 10^1 = 1 \times 10 = 10$$
$$4 \times 10^2 = 4 \times 100 = 400$$
$$3 \times 10^3 = 3 \times 1000 = 3000$$
$$3417$$

As seen in this example, each digit position has a value equal to the product of the digit appearing in the position and a corresponding power of 10.

Let us now briefly examine the binary, hexadecimal, and octal number systems.

9.2 BINARY NUMBERS

The binary number system is based on 2, rather than 10, so numbers are expressed as powers of 2. A short listing of powers of 2 is shown.

$$
\begin{aligned}
2^0 &= 1\\
2^1 &= 2\\
2^2 &= 4\\
2^3 &= 8\\
2^4 &= 16\\
2^5 &= 32\\
2^6 &= 64\\
2^7 &= 128\\
2^8 &= 256\\
2^9 &= 512\\
2^{10} &= 1024\\
2^{11} &= 2048\\
2^{12} &= 4096\\
2^{13} &= 8192.
\end{aligned}
$$

The value of a binary number in base 10 is determined by multiplying the value of each digit (0 or 1) by the corresponding power of two and summing all the products. The presence of a 1 in a digit position of a binary number indicates that the corresponding power of two is used in determining the value of the number. A 0 in a digit position indicates that the corresponding power of two is absent from the number. For example, the binary number 100111 may be expressed as

$$
\begin{aligned}
100111 &= (1 \times 2^5) + (0 \times 2^4) + (0 \times 2^3) + (1 \times 2^2)\\
&\quad + (1 \times 2^1) + (1 \times 2^0)\\
&= (1 \times 32) + (0 \times 16) + (0 \times 8) + (1 \times 4)\\
&\quad + (1 \times 2) + (1 \times 1)\\
&= 32 + 0 + 0 + 4 + 2 + 1\\
&= 39
\end{aligned}
$$

The binary number 100111 is equivalent to the decimal number 39.

To avoid confusion when several systems of notation are employed, it is customary to enclose each number in parentheses and to write the base as a subscript, in decimal notation. Thus, the previous example could be written as

$$(100111)_2 = (39)_{10}$$

Binary fractions are handled in the same way by assigning negative powers of two to the right of the binary point in ascending sequence. A short list of the negative powers of two follows:

$$
\begin{aligned}
2^{-1} &= \tfrac{1}{2} = .5 & 2^{-5} &= \tfrac{1}{32} = .03125\\
2^{-2} &= \tfrac{1}{4} = .25 & 2^{-6} &= \tfrac{1}{64} = .015625\\
2^{-3} &= \tfrac{1}{8} = .125 & 2^{-7} &= \tfrac{1}{128} = .0078125\\
2^{-4} &= \tfrac{1}{16} = .0625
\end{aligned}
$$

The binary fraction .1010 means

NUMBER THEORY TRIVIA

Occasional **hexadecimal palindromes** convert directly into numbers in base 10 that are also palindromes. For example

HEXADECIMAL PALINDROME	DECIMAL PALINDROME
17871	96369
189981	1612161

$$\begin{aligned}.1010 &= 1 \times 2^{-1} + 0 \times 2^{-2} + 1 \times 2^{-3} + 0 \times 2^{-4} \\ &= 1 \times 1/2 + 0 \times 1/4 + 1 \times 1/8 + 0 \times 1/16 \\ &= 1/2 + 0 + 1/8 + 0 \\ &= 4/8 + 1/8 \\ &= 5/8 \\ &= .625\end{aligned}$$

Thus,

$$(.1010)_2 = (.625)_{10}$$

As was the case with the binary number, an expansion of the binary fraction yields an equivalent decimal fraction.

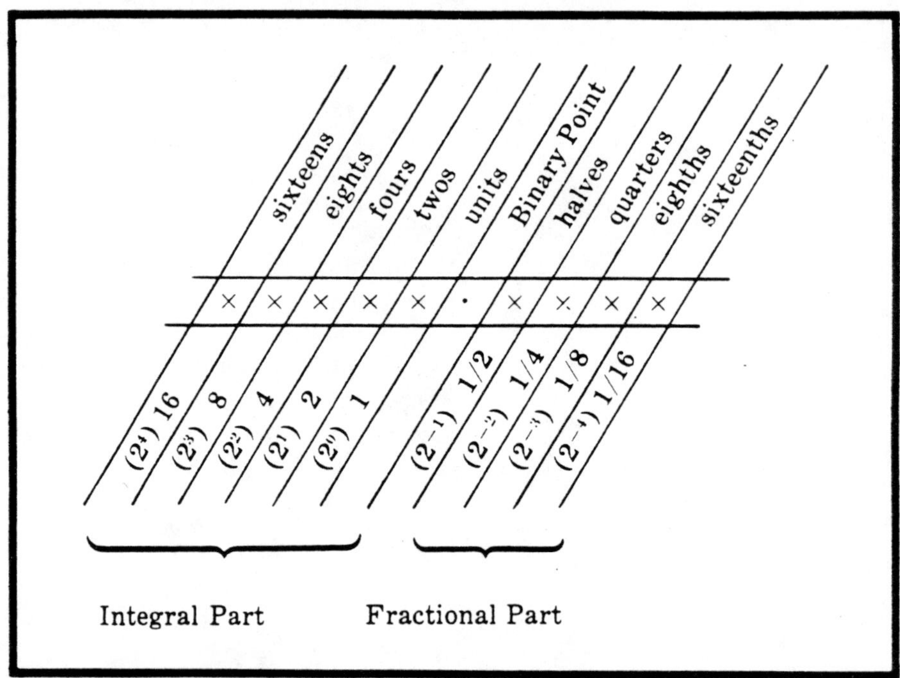

The positional weights of a binary number (powers of 2).

9.3 HEXADECIMAL NUMBERS

The hexadecimal number system is based on a radix of 16 and uses the following 16 digits: 0, 1, 2, 3, 4, 5, 6, 7, 8, 9, A, B, C, D, E, and F. The decimal value of a hexadecimal number is determined by multiplying the value of each digit by the corresponding power of 16 and summing all the products. For example, the equivalent decimal value of the hexadecimal number 83.5 may be determined in the following manner:

8 3 . 5

$5 \times 16^{-1} = 5 \times 1/16 = .3125$

$3 \times 16^{0} = 3 \times 1 = 3.$

$8 \times 16^{1} = 8 \times 16 = 128.$

131.3125

Thus, the hexadecimal number 83.5 is equivalent to the decimal number 131.3125, which is written

$$(83.5)_{16} = (131.3125)_{10}$$

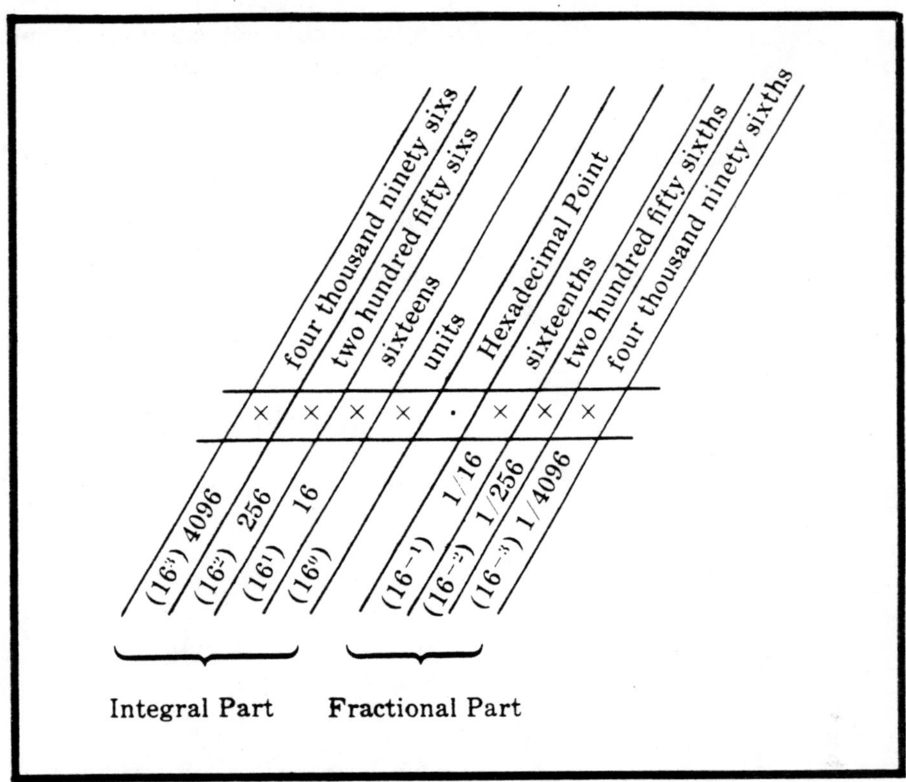

The positional weights of a hexadecimal number (powers of 16).

It may be useful to refer to the following for powers of 16 when converting hexadecimal numbers to decimal numbers:

Powers of Sixteen	Decimal Equivalent
16^4	65536
16^3	4096
16^2	256
16^1	16
16^0	1
16^{-1}	0.0625
16^{-2}	0.003906
16^{-3}	0.00024414

The hexadecimal number system provides a shortcut method of representing binary numbers. To convert a binary number to hexadecimal notation, divide the binary number into groups of four digits, starting from the binary point, and replace each group with the corresponding hexadecimal symbol. For example,

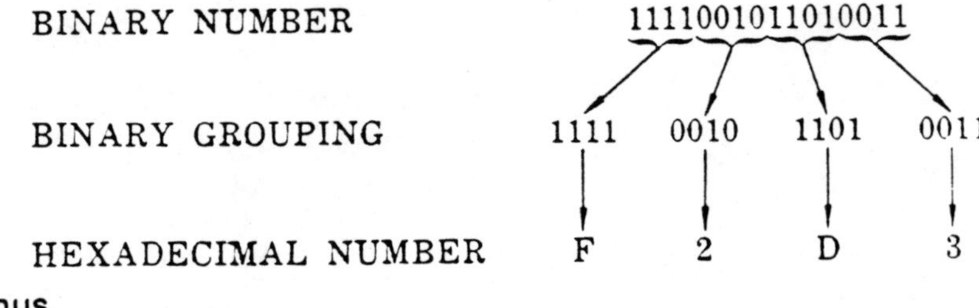

Thus,

$$(1111001011010011)_2 = (F2D3)_{16}$$

For reference, the binary grouping for each of the hexadecimal symbols is as follows:

Binary Grouping	Hexadecimal Symbol
0000	0
0001	1
0010	2
0011	3
0100	4
0101	5
0110	6
0111	7
1000	8
1001	9
1010	A
1011	B
1100	C
1101	D
1110	E
1111	F

Similarly, to convert a hexadecimal number into binary, substitute the corresponding group of four binary digits for each hexadecimal digit. For example,

$$\begin{aligned}(2A67.2F)_{16} &= 2 \quad A \quad 6 \quad 7 \;.\; 2 \quad F\\ &= 0010\ 1010\ 0110\ 0111\ .\ 0010\ 1111\\ &= (10101001100111.00101111)_2\end{aligned}$$

As shown in the previous example, leading zeros are discarded.

9.4 OCTAL NUMBERS

The octal number system is a number system whose base is eight and that uses the digits 0, 1, 2, 3, 4, 5, 6, and 7. Let us take the number $(143)_8$ and convert it into decimal:

$$\begin{aligned}(1\ 4\ 3)_8:\quad 3 \times 8^0 &= 3 \times 1 = 3\\ 4 \times 8^1 &= 4 \times 8 = 32\\ 1 \times 8^2 &= 1 \times 64 = 64\\ &\overline{(99)_{10}}\end{aligned}$$

20	51	44	5	33	76	67	22
32	77	66	23	15	54	41	10
1	50	55	14	26	63	72	37
27	62	73	36	4	45	60	11
46	3	12	57	61	30	35	74
64	25	40	71	47	2	13	56
53	16	7	42	100	31	24	65
75	34	21	70	52	17	6	43

An Order 8 magic square expressed in octal numbers.

Note that the conversion is done exactly as was done with binary and hexadecimal numbers, except that powers of eight were used. The following listing may be useful for such conversions:

Powers of Eight	Decimal Equivalent
8^7	2097152
8^6	262144
8^5	32768

8^4	4096
8^3	512
8^2	64
8^1	8
8^0	1
8^{-1}	0.125
8^{-2}	0.015625
8^{-3}	0.001953125
8^{-4}	0.000244140625

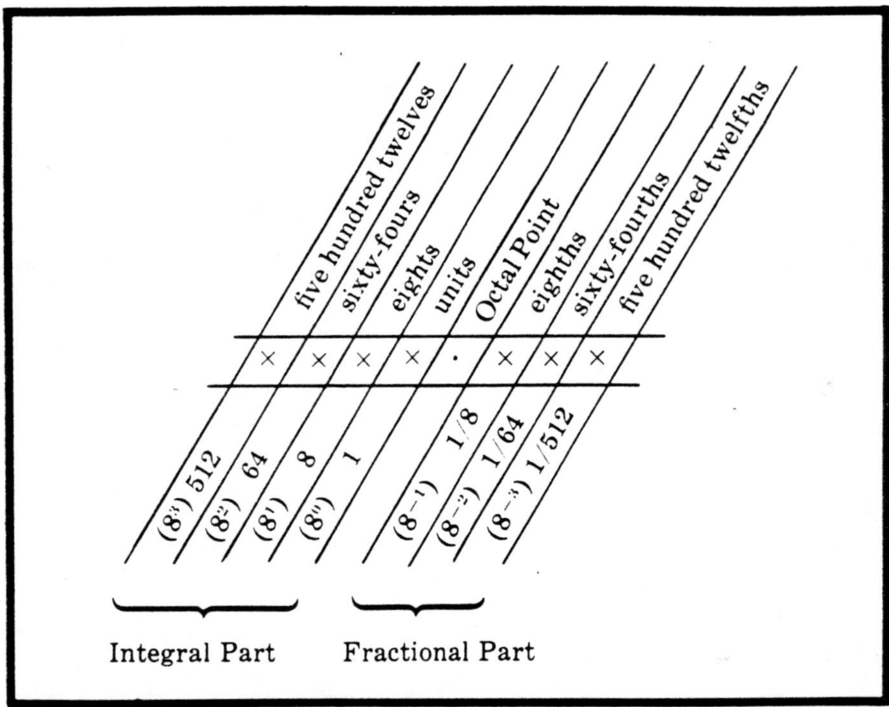

The positional weights of an octal number (powers of 8).

Similar to the hexadecimal number system, the octal number system has special characteristics that make it especially useful in many situations involving binary numbers. Since three binary digits may be grouped and represented by one octal digit, many binary numbers may be represented by using octal notation. This representation is useful when working with the operator's console of some computers and when printing the contents of a computer's memory. The following octal digits are used to represent the groupings of three binary digits:

Binary Grouping	Octal Digit
000	0
001	1
010	2
011	3
100	4
101	5
110	6
111	7

As was done with hexadecimal numbers, grouping of a binary number starts from the binary point and goes outward in both directions. For example,

$$(111010000001.110)_2 = 111\ 010\ 000\ 001 \cdot 110$$
$$= 7\quad 2\quad 0\quad 1 \cdot 6$$

Thus,

$$(111010000001.110)_2 = (7201.6)_8$$

Again, the writing and reading of binary numbers has been simplified in this case by using octal notation.

An octal number is converted to binary by using the opposite procedure. For example,

$$(263.4)_8 = 2\quad 6\quad 3\ .\ 4$$
$$= 010\ 110\ 011\ .\ 100$$
$$= (10110011.100)_2$$

9.5 NUMBER BASE CONVERSION PROGRAM

The decimal number 461 is interpreted as follows:

4	6	1
hundreds	tens	units

The octal number 715 is represented as:

7	1	5
sixty-fours	eights	units

These two numbers are the same because

$$7 \times 64 + 1 \times 8 + 5 = 461$$

A decimal number can be converted to another base by using the Pascal **div** and **mod** operators. To convert a number, reuse the **div** from each step; this is an iteration which is carried out until nothing is left.

461 **mod** 8 = 5	461 **div** 8 = 57
57 **mod** 8 = 1	57 **div** 8 = 7
7 **mod** 8 = 7	7 **div** 8 = 0(finished)

Here is a program which will convert a decimal number to base 8. Perhaps you would like to modify this program to convert a base 10 number into an equivalent base 16 number.

```
program NumberConversion;
(**** Converts a decimal integer to an ****)
(**** equivalent number in base 8 ****)
const
  Base = 8;
var
  Number      :Integer;(* Number in base 10 *)
  Digit       :array[1 . . 10] of Integer;
  I           :Integer;(* Index *)
```

NUMBER THEORY TRIVIA

Pairs of twin primes, when expressed in base 10, can only end in the digits 1, 3, 7, and 9; e.g., 179 - 181, 281 - 283, 827 - 829, 2999 - 3001. However, if primes are expressed in hexadecimal notation (base 16), there is no such limitation. The following table shows three pairs of 5-digit decimal twin primes with their hexadecimal equivalents.

DECIMAL	HEXADECIMAL
15269	3BA5
15271	3BA7
34649	8759
34651	875B
28349	6EBD
28351	6EBF

```
begin
   (**** Input a number in base 10 ****)
   write('ENTER AN INTEGER LESS THAN 32000: ');
   readln(Number);
   writeln('CONVERT ',Number,' IN BASE 10 TO BASE 8');
   writeln;
   (**** Initialize index ****)
   I := 1;
   (**** Iterate until no digits are left ****)
   repeat
      Digit[I] := Number mod Base;
      Number := Number div Base;
      I := I + 1;
   until Number = 0;
   (**** Print number in base 8 ****)
   write('CONVERTED NUMBER EQUALS ');
   (**** Reset index ****)
   I := I - 1;
   (**** Print digits of base 8 number ****)
   repeat
      write(Digit[I]);
      I := I - 1;
   until I < 1;
end.
```

COMPILE/RUN

```
ENTER AN INTEGER LESS THAN 32000: 1995
CONVERT 1995 IN BASE 10 TO BASE 8

CONVERTED NUMBER EQUALS 3713
```

NUMBER THEORY TRIVIA

Prime numbers that are palindromic both in decimal (base 10) and in hexadecimal (base 16) are extremely rare.

DECIMAL	HEXADECIMAL
2	2
3	3
5	5
7	7
11	11
353	161
787	313
94049	16F61
98689	18181

Can you write a program to produce additional primes of this form?

REVIEW EXERCISES

1. What is meant by the base of a number system?
2. Why do computers use the binary number system to represent numbers?
3. Match each of the following number system names (column 1) with its appropriate base (column 2):

1	2
(a) Decimal	(1) Base 16
(b) Binary	(2) Base 2
(c) Hexadecimal	(3) Base 8
(d) Octal	(4) Base 10

4. Express the positional values of the following decimal numbers:
   a. $(3,264)_{10}$
   b. $(48.9)_{10}$
   c. $(1,763.402)_{10}$
5. What is the decimal value of each of the following binary numbers?
   a. $(111010)_2$
   b. $(.1110111)_2$
   c. $(1010.0011)_2$
6. Represent the first twenty-two Fibonacci numbers as binary numbers.
7. Represent the current year as a binary number.
8. Represent your age as a hexadecimal number.
9. Expand the hexadecimal number $(2F.A6)_{16}$ into positional notation.
10. Express the hexadecimal number $(84.E)_{16}$ as a decimal value.
11. Show the binary equivalents of the following hexadecimal numbers:
    a. $(29)_{16}$
    b. $(42C)_{16}$
    c. $(63.4F)_{16}$
    d. $(163A7.2E7)_{16}$
12. Convert the following binary numbers to hexadecimal notation:
    a. $(11110010)_2$
    b. $(101100001110)_2$
    c. $(.11010010)_2$
    d. $(1011000000000011.01100111)_2$
13. Express $(237.4)_8$ as a decimal number.
14. Show the binary equivalents of the following octal numbers:
    a. $(36)_8$
    b. $(147)_8$

c. $(4.7)_8$
d. $(621.33)_8$

15. Convert the decimal number 186 to:
 a. a binary value
 b. a hexadecimal value.
 c. an octal value.

16. Convert the binary number 10111000001 to:
 a. a decimal value.
 b. an octal value
 c. a hexadecimal value

17. Convert the hexadecimal number 248 to:
 a. a binary value
 b. a decimal value
 c. an octal value

18. Use the number base conversion program to perform the following conversions:
 a. $(63F2)_{16} = (?)_{10}$
 b. $(463)_{10} = (?)_{16}$
 c. $(284)_{10} = (?)_2$
 d. $(691)_{10} = (?)_8$
 e. $(11001110)_2 = (?)_{16}$

19. Express the year 1989 in the octal number system - the scale of 8.

20. Express the year 2015 in the hexadecimal number system — the scale of 16.

I tend to agree with you - especially since 25(MOD 12) is my lucky number.

10
MODULAR ARITHMETIC

PREVIEW

A mathematical system is made up of two things: (1) a set of elements, and (2) one or more operations for combining the elements. A common mathematical system is the set of whole numbers, (0, 1, 2, 3, 4, 5, . . .) and the operation of addition. Here the set of elements is a set of numbers. Elements from this set are combined using the operation of addition. For example, the elements 4 and 7 are combined to get 11. Also, 8 and 14 are combined to get 22.

The operation of the mathematical system must be applicable to the set of the system. It would be meaningless to speak of the system made up of a set of rabbits and the operation of multiplication, since the operation has not been given meaning for the set of rabbits. We do not know how to multiply two rabbits. (Two rabbits may know how to multiply, but we don't know how to multiply rabbits.) We can only count the results. In this chapter we look at some of the abstractions in mathematics.

After you finish this chapter, you should be able to:

1. Identify modular arithmetic, clock arithmetic, the Chinese remainder theorem, and the ancient technique of "casting out nines."
2. Perform calculations in modular arithmetic.
3. 
See how a microcomputer can be programmed to work with clock arithmetic.
4. Use the casting out nines technique to check the result of an

NUMBER THEORY TRIVIA

There was a young fellow named Ben
Who could only count modulo ten
He said when I go
Past my last little toe
I shall have to start over again.

arithmetic operation.

5. See how a microcomputer can solve a number problem using the Chinese remainder theorem.

10.1 INTRODUCTION TO MODULAR ARITHMETIC

Modular arithmetic, sometimes called congruence arithmetic, modulo arithmetic, or clock arithmetic, is the application of fundamental operations that involve the use of numbers of one system only. Thus, the modulo M or mod M system uses only the numbers 0, 1, 2, . . ., (M − 1). The fundamental operations are the same as those of ordinary arithmetic except that if the number is greater than (M−1) it is divided by M and the remainder is used in place of the ordinary result. Generally speaking, when we write A modulo M, we mean the remainder when A is divided by M. More precisely, two numbers are considered congruent modulo M if they leave the same remainder. They leave the same remainder if their difference is divisible by M.

Example 1. 3 is congruent to 8(mod 5), since 8 − 3 = 5, and 5 is a multiple of 5.

Example 2. 3 is congruent to 53(mod 5), since 53 − 3 = 50, and 50 is a multiple of 5.

Example 3. 3 is not congruent to 19(mod 5) since 19 − 3 = 16, and 16 is not a multiple of 5.

Example 4. 23 is congruent to 8(mod 5), since 23 − 8 = 15, and 15 is a multiple of 5.

Example 5. $81 \equiv 0 \pmod{27}$, since 81 − 0 = 81, and 81 is a multiple of 27. The symbol ≡ means "is congruent to."

Another way of determining two congruent numbers mod m is to divide each by m and check the remainders. If the remainders are the same, then the numbers are congruent mod m. For example, 3/5 gives a remainder 3, and 53/5 gives a remainder 3; thus 3 is congruent to 53 (mod 5).

The following program will convert a specified positive integer into a mod M numeral.

```
program ModConversion;
var
  I      :Integer;(* Positive integer *)
  M      :Integer;(* Mod number *)
  A      :Integer;(* Number in Mod M *)
begin
  write('ENTER THE INTEGER ');
  readln(I);
```

```
    write('ENTER THE MOD ');
    readln(M);
    A := I - TRUNC(INT(I/M) * M);
    writeln('THE INTEGER IN MOD ',M,' IS ',A);
end.
```

The Pascal **mod** operator can be used to obtain the remainder after division of two integers. For example, 19 **mod** 5 gives a value of 4.

10.2 CLOCK ARITHMETIC

Modular arithmetic may seem new, but you will find you have been applying the principles involved for some time without having any mathematical language to describe what you were doing. A familiar instance of modular arithmetic is clock arithmetic. Any two hours differing by a multiple of 12 hours are given the same numeral, so we say that the hours are counted modulo 12. In fact, any periodicity gives rise to a congruence relation. Two calendar dates separated by a multiple of 7 fall on the same day of the week. Two angles differing by a multiple of 360° have common initial and terminal sides.

Let us use the clock face shown in Figure 10-1 and illustrate how clock arithmetic is performed. In this system 2 + 11 means start at 2, then move the hand 11 units (clockwise) to 1. In this arithmetic, 2 + 11 = 1.

Figure 10-1
Clock arithmetic - 2 + 11 = 1.

NUMBER THEORY TRIVIA

It is India that gave us the ingenious method of expressing all numbers by means of ten symbols, each symbol receiving a value of position as well as an absolute value; a profound and important idea which appears so simple to us now that we ignore its true merit.

Pierre Laplace
(1749-1827)
French mathematician

If a scientist is performing an experiment in which it is necessary to keep track of the total number of hours elapsed since the start of the experiment, he may label the hours sequentially, as 1, 2, 3, etc. When 53 hours have elapsed, it is 53 o'clock experiment time. How does he reduce experiment time to ordinary time? If zero hours experiment time corresponds to midnight, his task is easy. He simply divides by 12 and the remainder is the time of day. For example, 53 experimental time is 5 o'clock, because 12 goes into 53 with a remainder of 5; 53 is congruent to 5 modulo 12.

Example 1. Find the clock equivalent of 567. The answer is 3, since a remainder of 3 is obtained when 567 is divided by 12. The fact that 12 goes into 567 forty-seven times has no significance here.

Example 2. Find the clock equivalent of 180. The answer is 0, since a remainder of 0 is obtained when 180 is divided by 12. 180 hours would involve fifteen complete rotations about the clock. The hand would come to rest at 0.

Example 3 Find the clock equivalent of −3. Minus 3 corresponds to turning the hand 3 hours in the opposite or counter-clockwise direction. The final position of the hand is at 9. Therefore, the equivalent of −3 is 9.

Example 4. 12 − 3 = 9

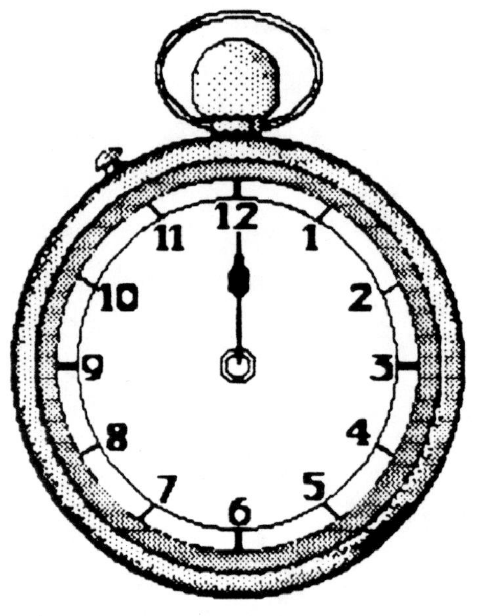

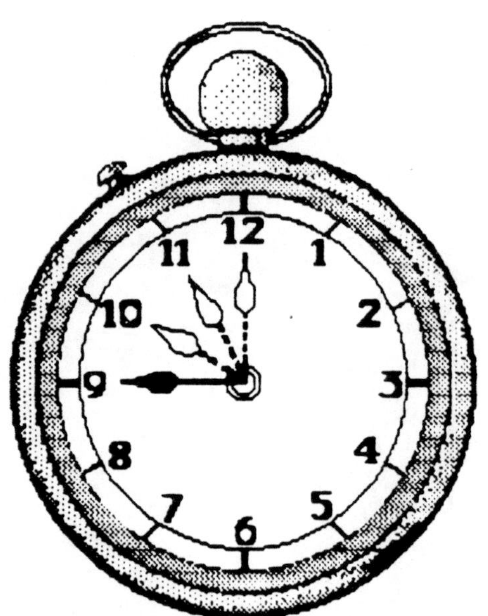

Example 5. 5 + 12 = 5

Example 6. 7 + 5 = 12

Example 7. 4 + 11 = 3

Example 8. 9 − 5 = 4

Example 9. 9 + 4 = 1

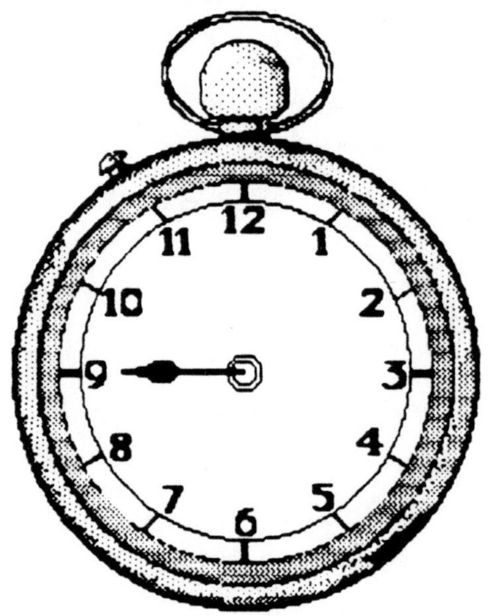

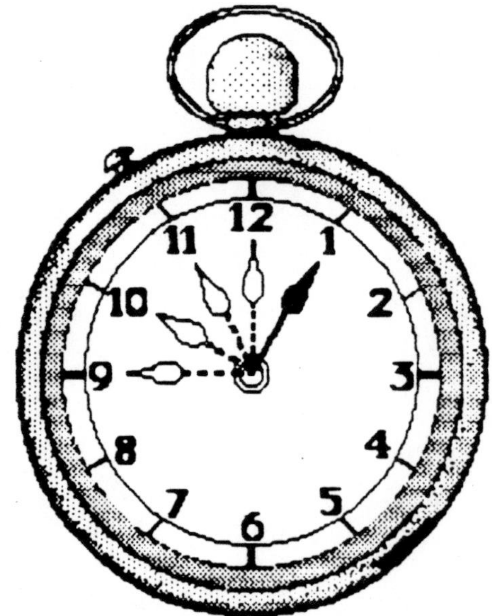

The following program starts at random times and adds random number of hours. The random times must be numbers from 1 to 12. The random numbers of hours could have been from virtually any range, but 1 to 36 is used in this program. The program picks eight pairs of random numbers, with T for time and H for hours, and adds them to see whether the sum is less than or equal to 12. If the sum is less than or equal to 12, the program prints the sum as the time. If the sum is greater than 12, the program determines and prints the number modulo 12. After a sum is printed, the program loops back to pick another pair of numbers, and repeats the process until 8 pairs of numbers have been picked and processed.

```
program ClockArithmetic;
var
  T,H,S,L        :Integer;
begin
  for L := 1 to 8 do
    begin
      T := RANDOM(12) + 1;
      H := RANDOM(36) + 1;
      S := (T + H) mod 12;
      if S = 0 then S := 12;
      writeln(H:2, ' HOURS FROM ', T, ' O''CLOCK');
      writeln('IT WILL BE ', S, 'O''CLOCK');
      writeln;
    end;
end.
```

NUMBER THEORY TRIVIA

Prime quadruples are prime number sequences such as

101 103 107 109
and
191 193 197 199

Both of these sequences fit the pattern

n n+2 n+6 n+8

Can you write a program to generate several prime quadruples?

```
COMPILE/RUN

8 HOURS FROM 8 O'CLOCK
  IT WILL BE 4 O'CLOCK

32 HOURS FROM 8 O'CLOCK
  IT WILL BE 4 O'CLOCK

31 HOURS FROM 9 O'CLOCK
  IT WILL BE 4 O'CLOCK

33 HOURS FROM 1 O'CLOCK
  IT WILL BE 10 O'CLOCK

5 HOURS FROM 2 O'CLOCK
  IT WILL BE 7 O'CLOCK

18 HOURS FROM 9 O'CLOCK
  IT WILL BE 3 O'CLOCK

4 HOURS FROM 10 O'CLOCK
  IT WILL BE 2 O'CLOCK

22 HOURS FROM 2 O'CLOCK
  IT WILL BE 12 O'CLOCK
```

10.3 CASTING OUT NINES TECHNIQUE

An ancient technique known as casting out nines is a method used to check multiplication or addition. It is based on the excess of nines in the digits of an integer. The excess of nines is the remainder when the sum of the digits is divided by 9. For example, the excess of nines in 9357 is 6, since $9 + 3 + 5 + 7 = 24$, and the remainder when 24 is divided by 9 is 6. The sum of several numbers can be checked as follows:

$ 14.32	$1 + 4 + 3 + 2 \equiv 1 \pmod 9$
1.48	$1 + 4 + 8 \equiv 4 \pmod 9$
76.25	$7 + 6 + 2 + 5 \equiv 2 \pmod 9$
75.65	$7 + 5 + 6 + 5 \equiv 5 \pmod 9$
4.13	$4 + 1 + 3 \equiv 8 \pmod 9$
$171.83	$1 + 7 + 1 + 8 + 3 \equiv \overline{2} \pmod 9$

This method can be applied to any arithmetical expression involving any combination of additions and multiplications. The casting out nines technique is based on congruence properties modulo 9.

This same idea can be used to test a large number for divisibility by 9. Is 5438777 evenly divisible by 9?

$$5438777 = 5 + 4 + 3 + 8 + 7 + 7 + 7$$
$$\equiv 5 \not\equiv 0 \pmod 9,$$

therefore, 5438777 is not evenly divisible by nine.

NUMBER THEORY TRIVIA

There are nine prime numbers between 9,999,900 and 10,000,000:

9,999,901
9,999,907
9,999,929
9,999,931
9,999,937
9,999,943
9,999,971
9,999,973
9,999,991

However, there are only two primes among the next hundred integers:

10,000,019
10,000,079

NUMBER THEORY TRIVIA

There are two different ways of expressing the prime number 809 as the sum of smaller primes with no digit used more than once, on either side of the equation.

809 = 761 + 43 + 5
809 = 743 + 61 + 5

The casting out nines technique can also be used to form pretended mind reading tricks you can use to entertain your friends.

Begin by asking your friend to write down any number, containing as many digits as he or she wishes, such as 6347.

$$\begin{array}{r} 6347 \\ 7436 \\ \hline 1089 \end{array}$$

Then tell him or her to write the same number backward (7436). Then tell him to subtract the smaller from the larger (yielding 1089). Next tell him or her to cross out any one digit of his or her choice in this number except that it must not be a zero. Then ask him or her to read you the remaining digits, and you will tell him or her the digit he or she crossed out. (In this example, if he or she crosses out the 8, he or she then reads you the digits "109." You immediately tell him or her: "You crossed out an 8").

The trick is performed as follows: when you hear the remaining digits, you mentally add them together (1 × 0 × 9 = 10). If the result contains more than one digit, add them together (1 + 0 = 1). Continue until you have only a single digit (1 in this case), then subtract it from 9, (9 − 1 = 8). The answer is the digit crossed out. An exception occurs when the sum of the digits is 9. In this case the crossed out digit is the sum itself, 9.

There are many interesting variations of this demonstration. For example, you can ask your friend to write any number containing any number of digits, then to write the sum of the digits below and subtract, then to cross out any one of the nonzero digits in the result, as shown:

$$\begin{array}{r} 763218 \\ -27 \\ \hline 76\not{3}191 \end{array}$$

Another variation is to give your friend these instructions:

1. Write down any positive three-digit integer in which the first digit differs from the third by at least two, 265 for example.
2. Now write the same number backwards, below the original number (562).
3. Subtract the smaller of these numbers from the larger (297).
4. Reverse the result (792), and add it to the number you had before you reversed it (792 + 297 = 1089).

NUMBER THEORY TRIVIA

A Chinese remainder theorem example. Ask anyone to select a number less than 60. Request the person to perform the following operations: (1) To divide it by 3 and mention the remainder; suppose it to be *a*. (2) To divide it by 4, and mention the remainder; suppose it to be *b*. (3) To divide it by 5, and mention the remainder; suppose it to be *c*. Then, the number selected is the remainder obtained by dividing 40*a* + 45*b* + 36*c* by 60.

The example shows the steps for 265.

```
 265
 562
----
 297
 792
----
1089
```

At this point you can ask your friend to be quiet while you meditate to discover his final result by ESP. After a minute or so, you announce the answer: 1089. This will always be correct if he has carried out the steps correctly, no matter what number he started with.

10.4 CHINESE REMAINDER THEOREM

The Chinese remainder theorem states the following result: given a set of congruences of the form

X is congruent to A (mod M)

where each pair of M's is relatively prime (that is, it has no factor greater than 1 in common), if we let P be the product of the M's, we then have a unique value B, $0 \leq B < P$ whereby every X having all the given properties is congruent to B modulo P.

This theorem is used in an old mind reading trick when the huckster asks someone in the audience to think of a number between one and 30. Then his spiel goes: "Don't tell me the number, but divide the number by two and give me the remainder, next divide the number by five, and give me the remainder." When the huckster has the three remainders he is able to calculate the original number.

Take the number 12, for example:

12/2 = 6 Remainder 0
12/3 = 4 Remainder 0
12/5 = 2 Remainder 2

Given these three remainders, it is easy to determine the original number since there is only one number less than 30 with these three remainders. Try it for yourself. Note that the three divisors are relatively prime (no common factors within the range 2 × 3 × 5 = 30).

The following program uses the Chinese remainder theorem with the microcomputer playing the part of the huckster. In the program 5, 7, 9 are used for the divisors (Range = 5 × 7 × 9 = 315).

```
program ChineseRemainderTheorem;
var
  R5      :Integer;(* Remainder of number/5 *)
  R7      :Integer;(* Remainder of number/7 *)
  R9      :Integer;(* Remainder of number/9 *)
  A       :Integer;(* Calculation variable *)
  X       :Integer;(* Computed number *)
begin
  writeln('I WANT YOU TO THINK OF A NUMBER');
  writeln('LESS THAN 316. WRITE THIS NUMBER');
  writeln('DOWN AND DIVIDE BY 5. NOW ENTER');
  write('THE REMAINDER LEFT OVER: ');
  readln(R5);
  writeln;
  writeln('NOW DIVIDE YOUR ORIGINAL NUMBER BY');
  write('7 AND ENTER THIS REMAINDER: ');
  readln(R7);
  writeln;
  writeln('NOW DIVIDE YOUR ORIGINAL NUMBER BY');
  write('9 AND ENTER THIS REMAINDER: ');
  readln(R9);
  writeln;
  (**** Calculate number ****)
  A := 126 * R5 + 225 * R7 + 280 * R9;
  X := A - TRUNC(A/315) * 315;
  writeln('I AM HAPPY TO TELL YOU THAT YOUR');
  writeln('NUMBER CHOSEN WAS ',X);
end.
```

COMPILE/RUN

```
I WANT YOU TO THINK OF A NUMBER
LESS THAN 316. WRITE THIS NUMBER
DOWN AND DIVIDE BY 5. NOW ENTER
THE REMAINDER LEFT OVER: 0

NOW DIVIDE YOUR ORIGINAL NUMBER BY
7 AND ENTER THIS REMAINDER: 2

NOW DIVIDE YOUR ORIGINAL NUMBER BY
9 AND ENTER THIS REMAINDER: 1

I AM HAPPY TO TELL YOU THAT YOUR
NUMBER CHOSEN WAS 100
```

今有物不知其數三三數之賸二五五數之賸
三七七數之賸二問物幾何

荅曰二十三
術曰三三數之賸二置一百四十五五數
之賸三置六十三七七數之賸二置三十
并之得二百三十三以二百一十減之即
得凡三三數之賸一則置七十五五數之
賸一則置二十一七七數之賸一則置十
五一百六以上以一百五減之即得

The earliest known formulation of the Chinese remainder theorem appears to be in the *Sun Tzu Suan-ching* (i.e., the mathematical classic of Sun Tzu) which has been dated between 280 A.D. and 473 A.D.

In the previous example, the number 100 was chosen. The algorithm for calculating the number is:

$$X = (126 \times R5 + 225 \times R7 + 280 \times R9) \bmod 315$$
$$X = (26 \times 0 + 225 \times 2 + 280 \times 1) \bmod 315$$
$$X = (730) \bmod 315$$
$$X = 100$$

NUMBER THEORY TRIVIA

The largest perfect square whose digits are in strictly decreasing order is

$$961 = 31^2$$

Can you write a program to find the only other perfect square of this type?

X is the number. R5 is the remainder when X is divided by 5, R7 is the remainder when X is divided by 7, and R9 is the remainder when X is divided by 9. Mod 315 indicates remainder arithmetic, where the parenthesized quantity is divided by 315 but the result is the remainder, not the quotient.

The algorithm in the program is not exactly of this form since BASIC does not have a modulus function. However, BASIC does have an INTEGER function, which is used in the program (line number 340) to do the remainder arithmetic.

REVIEW EXERCISES

1. Define precisely the concept of "congruence modulo m."
2. Which of the following are true?
   a. 126 is congruent to 1(mod 7)
   b. 144 is congruent to 12(mod 144)
   c. 1987 is congruent to 0(mod 1987)
   d. 246 is congruent to 150(mod 6)
3. Which of the following are true?
   a. 47 is congruent to 2(mod 5)
   b. 5670 is congruent to 270(mod 365)
   c. 108 is congruent to 12(mod 8)
   d. 2001 is congruent to 39(mod 73)
4. Use the program in Section 10.1 to find x.
   a. 5872 is congruent to x(mod 119)
   b. 5827300 is congruent to x(mod 365)
   c. 6214002 is congruent to x(mod 4230)
5. Using the twelve-hour clock arithmetic system, find the equivalent of the following numbers. In each case, your answer should be one of the numbers 0, 1, 2, 3, 4, 5, 6, 7, 8, 9, 10, 11.
   a. 33 c. 15 e. 126
   b. 1987 d. −36 f. −24
6. Using the twelve-hour clock arithmetic system, evaluate the following:
   a. 2 + 3 d. 10 + 11
   b. 9 + 9 e. 9 + 4
   c. 8 − 5 f. 1 − 10
7. Perform the indicated operations using arithmetic for a 12-hour clock.
   a. 9 + 6 d. 2 + 7
   b. 7 + 10 e. 6 + 3
   c. 5 + 7 f. 4 + 8
8. Your doctor tells you to take a certain medication every 8 hours. If you begin at 8:00 A.M., show that you will not have to take the medication between midnight and 7:00 A.M.
9. What is 3 − 7 on the 12-hour clock? Note that 3 − 7 = T is equivalent to T + 7 = 3.

NUMBER THEORY TRIVIA

The sum of the first n even numbers is $n \times (n + 1)$. For example

$2 + 4 + 6 = 12$
$2 + 4 + 6 + 8 + 10 = 30$
$2 + 4 + 6 + 8 + 10 + 12 = 42$

10. Make a complete table of addition facts on the 12-hour clock.
11. Solve the equation T + 6 = 2 for T where T may be replaced by any one of the numerals on a 12-hour clock.
12. Solve each equation where T may be replaced by any one of the numerals on a 12-hour clock:
    a. T - 3 = 11
    b. 3 + T = 2
    c. T + 7 = 5
13. As shown in Section 10.3, a number is divisible by 9 only if the sum of its digits is divisible by 9. Determine if 234,648 is divisible by 9.
14. Write a program to determine if the number 39,827,437 is divisible by 9. The program should use the casting out nines technique.
15. Check the addition of the following numbers using the casting out nines technique.

```
 14,745
 23,610
 10,100
 21,007
  6,143
 12,841
 ------
 88,446
```

16. Write a program to print out a calendar in a familiar form in response to an integer representing the year (e.g., 1990). To calculate the day of a week, use the following formula:

$$F = [(2.6M - 0.2) + K + D + (D/4) + (C/4) - 2C] \bmod 7$$

where brackets represent the integer part, and mod 7 implies the remainder after division by 7. Variables are defined as follows:

F = day of week (Sunday = 0, Monday = 1, . . .)
K = day of month (1, 2, 3, . . ., 31)
C = century (18, 19, 20, . . .)
D = year in century (1, 2, 3, . . ., 99)
M = month number with January and February taken as months 11 and 12 of the preceding year (March = 1, April = 2,. . .). Thus, May 14, 1989 has K = 14, C = 19, D = 89, M = 3. Note: This formula applies only to years after the calendar change in 1752.

It says our chances of being rescued is less than finding a proof for Fermat's Last Theorem.

11

POTPOURRI

PREVIEW

In previous chapters, we considered a wide variety of number theory problems, with each chapter devoted to problems relating to a particular area of number theory. But we have only scratched the surface of the many topics that are included in number theory. It is not possible to include all of this material in any one book; but presented here is a sampling of some other number theory topics. The chapter is what its name says – a potpourri.

After you complete this chapter, you should be able to do the following:

1. Identify Diophantine problems, map-coloring problems, Pythagorean triples, Pascal's triangle and Palindrome numbers.
2. Devise number tricks, and solutions to famous problems such as the "sailors and coconuts" problem and the "crossing the river" riddle.
3. See how a microcomputer can be used to produce Pythagorean triples and Pascal's triangle.
4. See how a microcomputer can be used to determine if a number is a palindrome.
5. Understand two famous unsolved problems in the field of number theory: Fermat's last theorem and Goldbach's conjecture.

NUMBER THEORY TRIVIA

The following squares of palindromes contain nine decimal digits exactly once each.

$35853^2 = 1285437609$
$54918^2 = 3015986724$
$84648^2 = 7165283904$

Can you write a program to produce other palindromes of this form?

11.1 MIND READING TRICKS

This section contains several tricks that are based on fairly simple mathematical operations. You may wish to let the computer become the mind reader by programming one or more of the examples.

* * * **Trick 1** * * *

The mind reader (computer) asks a person in his audience to think of a number, multiply it by 5, add 6, multiply by 4, add 9, multiply by 5, and state the result.

The person chooses the number 12, calculates successively 60, 66, 264, 273, 1365, and announces the last number.

The mind reader (computer) subtracts 165 from the result, gets 1200, knocks off the two zeros, and informs the person that 12 was his original number.

The trick is easily seen if put in mathematical symbols. If the number chosen is a, then the successive operations yield

$$5a$$
$$5a + 6$$
$$20a + 24$$
$$20a + 33 \text{ and}$$
$$100a + 165$$

When the mind reader (computer) is told this number, it is evident that he can determine a if he subtracts 165 and then divides by 100.

* * * **Trick 2** * * *

If the mind reader (computer) desires to tell a person the result without asking any questions, he must arrange the various operations so that the original number drops out. Here is an example in which three unknown numbers are introduced and eliminated.

The mind reader (computer) says: Think of a number. Add 10. Multiply by 2. Add the amount of change in your pocket. Multiply by 4. Add 20. Add 4 times your age in years. Divide by 2. Subtract twice the amount of change in your pocket. Subtract 10. Divide by 2. Subtract your age in years. Divide by 2. Subtract your original number.

The person, who chooses the number 7, has 30 cents in his pocket, and is 20 years old, thinks: 7, 17, 34, 64, 256, 276, 356, 178, 118, 108, 54, 34, 17, 10.

The mind reader (computer) says your result is 10, is it not? The person replies, "Right!"

NUMBER THEORY TRIVIA

The sum of the first *n* odd numbers is $n \times n$. For example, the sum of the first 10 numbers is 10×10.

$1 + 2 + 3 + \ldots + 17 + 19 = 100$

The sum of the first 60 odd numbers is 60×60. The sum of the first 500 odd numbers is 500×500.

In this case, if we denote the person's original number by a, the amount of change in his pocket by b, and his age in years by c, the successive operations give

$$
\begin{array}{r}
a \\
a + 10 \\
2a + 20 \\
2a + 20 + b \\
8a + 80 + 4b \\
8a + 100 + 4b + 4c \\
4a + 50 + 2b + 2c \\
4a + 50 + 2c \\
4a + 40 + 2c \\
2a + 20 + c \\
2a + 20 \\
a + 10 \\
10
\end{array}
$$

Problems of this type can be set up in many ways.

* * * Trick 3 * * *

Many tricks of the kind we are discussing are based upon the principle of positional notation. Consider the following.

The mind reader (computer) says: throw three dice and note the three numbers which appear. Operate on these numbers as follows: Multiply the number on the first die by 2, add 5, multiply by 5, add the number on the second die, multiply by 10, add the number on the third die, and state the result.

The person throws a 2, 3, and a 4, and thinks: 4, 9, 45, 48, 480, 484. He gives the answer 484.

The mind reader (computer) subtracts 250 and gets 234. He then states that the numbers thrown were 2, 3, and 4.

* * * Trick 4 * * *

The mind reader (computer) says: choose any prime number greater than 3, square it, add 17, divide by 12, and remember the remainder.

The person thinks: 11, 121, 138, 116/12, 6.

The mind reader (computer) states that the remainder is 6.

Here use is made of the fact that any prime number greater than 3 is of the form $6n \pm 1$, where n is a whole number. The symbol $\pm$ means plus or minus, its square is then of the form $36n^2 \pm 12n + 1$.

NUMBER THEORY TRIVIA

An infinite number of palindromic integers generate new palindromes when cubed, e.g.,

$101^3 = 1030301$

This number, when divided by 12, leaves a remainder of 1. Now the mind reader (computer) had the person add 17, which, divided by 12, leaves a remainder of 5. The final remainder must thus be 1 + 5, or 6.

The mind reader (computer) can vary this trick by using a number other than 17. The number when divided by 12, will have a remainder equal to k. Then the final remainder will always be 1 + k.

11.2 SAILORS AND COCONUTS

You have probably already encountered the sailors and coconuts problem that has tantalized many a problem solver.

Five sailors were stranded on a desert island where the only food they could find was coconuts. These were gathered all day, and by nightfall the men were so tired that, rather than eating, they decided to wait till the next day to split up the coconut feast. The most suspicious member of the lot arose after he was certain his fellows were asleep and divided the pile of coconuts into five equal shares, took his share and hid it. To conceal his act he pushed the other shares back into one large pile. But lo, a single coconut remained. This one he gave to the only other being the men had found on the island, a monkey. One by one, (in order of how suspicious each was), the other men arose and repeated the actions of the first sailor, each giving a single remaining coconut to the monkey. (The monkey learned quite quickly to stay around as each sailor arose).

Upon arising the next day the pile of coconuts was divided equally with one remaining coconut going to the monkey. The following program finds out how many coconuts there were and how many each sailor got.

```
program SailorsAndCoconuts;
var
  A,S,K,N      :Integer;
begin
  (**** Set number of sailors ****)
  S := 5;
  A := 1;
  K := A;
  N := 0;
  repeat
    while(S*K+1) / (S-1) <> (S*K+1) div (S-1) do
      begin
        A := A + 1;
        K := A;
        N := 0;
      end;
    N := N + 1;
    K := (S * K + 1) div (S - 1);
  until N = S;
  writeln('THERE WERE ', S*K+1, ' COCONUTS');
  writeln('IN THE MORNING EACH SAILOR GETS ', A, ' COCONUTS');
end.
```

COMPILE/RUN

```
THERE WERE 15621 COCONUTS
IN THE MORNING EACH SAILOR GETS 1023 COCONUTS
```

11.3 DIOPHANTINE PROBLEMS

If you are interested in puzzles, you are probably familiar with problems of the following type. A student's transcript shows T 3-hour courses and F 5-hour courses, for a total of 64 hours. Find T and F. From the description of the problem we get the equation

$$3T + 5F = 64.$$

This one equation, in two unknowns, constitutes the entire translation into algebra of the conditions. However, there is a difference between this problem and those considered in previous chapters. Here we want only integral solutions. To solve this problem experimentally, note that F must be less than 13, no matter what T is, since $5 \times 13 = 65$, and 65 is greater than 64. If $F = 12$ we get $3T = 4$, which is not a solution. Continuing to decrease F, we find four possible solution pairs, corresponding to the four points on the line $3T + 5F = 64$ in Figure 11-1.

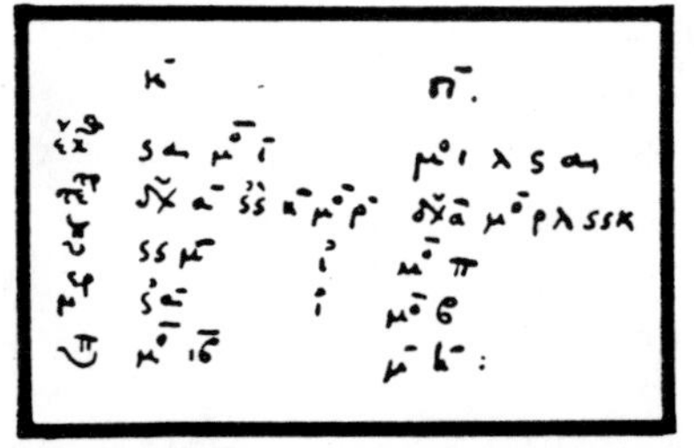

From a manuscript of the 14th century showing symbolism then in use. The problem is to find two numbers such that their sum is equal to 20 and the difference of their squares to 80.

Equations like the previous one that have to be solved for integral values of the unknowns are called Diophantine equations, after the Greek mathematician Diophantus. Frequently such an equation has no solution. For example, $6X + 9Y = 16$ has no solution in integers, since the lefthand side is divisible by 3 and the righthand side is not.

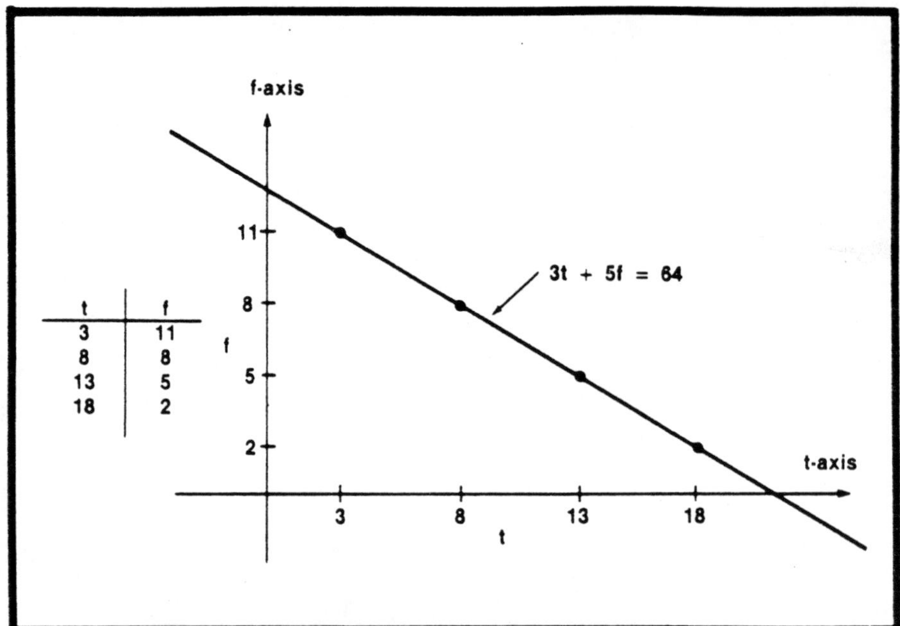

Figure 11-1
Experimental solution of 3t + 5f = 64.

t	f
3	11
8	8
13	5
18	2

11.4 MAP COLORING PROBLEMS

When coloring a geographical map, it is customary to give different colors or different shades to any two countries that have a portion of their boundary in common. It has been found empirically that any map, no matter how many countries it contains or how they are situated, can be so colored by using only four different colors. It is easy to see that no smaller number of colors will suffice for all cases (see Figure 11-2).

The four color problem is one of the most celebrated challenges in mathematics. It is of great intellectual interest and has intrigued many people from all paths of life. Its solution has little or nothing whatsoever to do with making maps. A map maker is and always will be able to print maps using as many different colors as he needs.

No one has ever been able to produce a map that would require more than four colors, and until 1976 no one had been able to prove that four colors are sufficient for all maps. The computer-aided proof of the century-old conjecture was completed in 1976 and published in 1977, by Kenneth Appel and Wolfgang Haken at the University of Illinois. This proof cannot in fact be checked by hand calculations. In more than a century of research no simple elegant proof of this problem has ever been demonstrated. But the search for such a proof has stimulated development of new branches of mathematical science in the fields of combinatorial mathematics and topology.

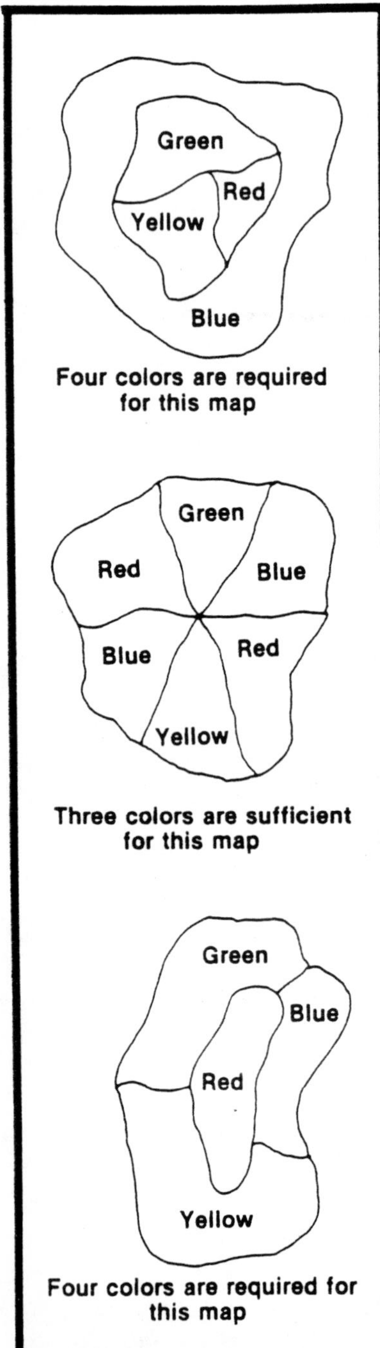

Figure 11-2
Map coloring.

Two amazing revelations appeared in the 1977 papers of Appel and Haken. The first was that these two mathematicians had finally solved the problem that had frustrated all attempts for one and a quarter centuries. The second was even more significant in its

NUMBER THEORY TRIVIA

Two young braves and three squaws are sitting proudly side by side. The first squaw sits on a buffalo skin with her 50 pound son. The second squaw is on a deer skin with her 70 pound son. The third squaw, who weighs 120 pounds, is on a hippopotamus skin. Therefore, the squaw on the hippopotamus is equal to the sons of the squaws on the other two hides.

implications for the world of mathematics: the successful proof was dependent upon the use of a highspeed computer and could not have been accomplished without it. The computer-assisted proof of the four-color conjecture demanded about 10 billion logical decisions and required more than 1200 hours of work analyzing thousands of configurations by computer.

Pythagoras (550-500 B.C.) was born on the Aegean island of Samos. After travelling in Egypt and elsewhere, he migrated to the Greek seaport of Crotona (now in Italy) where he founded the Pythagorean School. This was a closely-knit brotherhood whose purpose was to further the study of mathematics. Pythagoras has rightly been called one of the founding fathers of the science of mathematics. His most famous rule or theorem is what he discovered about right angled triangles.

11.5 PYTHAGOREAN TRIPLES

An early number theory is the Pythagorean problem. As you know, in a right-angled triangle, the lengths of the sides satisfy the Pythagorean relation

$$A^2 + B^2 = C^2$$

where C is the length of the hypotenuse, the side opposite the right angle.

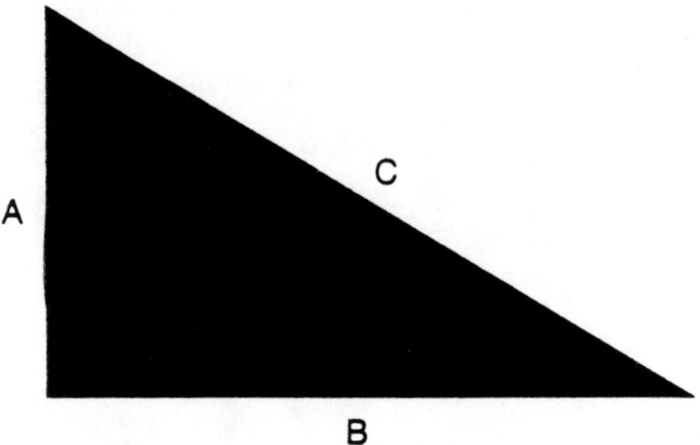

This equation makes it possible to compute the length of one side of a right triangle when you know the other two.

Sometimes all the side lengths A, B, C (reference diagram) are integers. The simplest case is A = 3, B = 4, and C = 5. There are many other cases of integral solutions of the Pythagorean equation:

A	B	C
7	24	25
8	15	17
9	40	41
11	60	61
12	35	37
13	84	85
20	21	29
28	45	53
33	56	65

These sets of three integers, A, B, and C, are called Pythagorean triples.

Shown above is a Moslem manuscript, written in 1258, which shows the Pythagorean theorem.

Pythagorean triples may be generated from the following formulas

$$A = R^2 - S^2$$
$$B = 2RS$$
$$C = R^2 + S^2$$

where R and S are relatively prime, where $R > S > 0$, and where one of the numbers R and S is even and the other is odd.

NUMBER THEORY TRIVIA

A Pythagorean triangle represented by the equation

$$9999^2 + 137532^2 = 137895^2$$

has an area of 687591234, which contains all nine non-zero digits, whereas one of the sides contains only a single repeated digit.

Can you write a program to find another Pythagorean triangle whose area contains all nine digits and one of whose sides contains only a single repeated digit?

The microcomputer can be used to eliminate the drudgery of calculating Pythagorean triples. The following Pascal program produces Pythagorean triples. Run the program using an N of 50. By increasing the value of N you can produce additional Pythagorean triples.

```
program PythagoreanTriples;
var
  A,B,C,AA,BB,CC,N        :Integer;
begin
  write('ENTER N: ');
  readln(N);
  for A := 1 to N do
    begin
      AA := A * A;
      for B := A to N do
        begin
          BB := AA + B * B;
          C := B + 1;
          repeat
            CC := BB - C * C;
            if CC = 0 then writeln(A:6, B:10, C:10);
            C := C + 1;
          until CC < 0;
        end;
    end;
end.
```

Note: Program output is shown on page 214

This postage stamp issued by Nicaragua shows the Law of Pythagoras, $A^2 + B^2 = C^2$.

COMPILE/RUN

ENTER N: 70

3	4	5
5	12	13
6	8	10
7	24	25
8	15	17
9	12	15
9	40	41
10	24	26
12	16	20
12	35	37
14	48	50
15	20	25
15	36	39
16	30	34
18	24	30
20	21	29
20	48	52
21	28	35
24	32	40
24	45	51
24	70	74
25	60	65
27	36	45
28	45	53
30	40	50
32	60	68
33	44	55
33	56	65
36	48	60
39	52	65
40	42	58
42	56	70
45	60	75
48	55	73
48	64	80
48	90	102
51	68	85
54	72	90
56	90	106
56	105	119
57	76	95
60	63	87
60	80	100
60	91	109
63	84	105
65	72	97
66	88	110

This Greek postage stamp illustrates the Pythagorean Theorem with a sort of tile pattern, suggesting a simple way to demonstrate the theorem. Notice that the square on the longest side of the triangle contains 25 small squares, that the squares on the other two sides contain 16 squares and 9 squares, and that 25 = 16 + 9.

NUMBER THEORY TRIVIA

A Pythagorean triple, or Pythagorean triangle (PT), consists of 3 positive integers a, b, c, such that $a^2 + b^2 + c^2$, where a and b correspond to the arms, and c to the hypotenuse of a right triangle. In a primitive PT, these integers are coprime, that is, they have no common factor. A **prime PT** is a primitive PT where the odd sides, a and c, are prime. The following table lists several prime PTs:

a	b	c
3	4	5
5	12	13
11	60	61
19	180	181
29	420	421
59	1740	1741
61	1860	1861
71	2520	2521
79	3120	3121
101	5100	5101
131	8580	8581
139	9660	9661

Can you write a program to generate all prime PTs up to a hypotenuse of at least 30,000?

Look at the list of triples produced by the program. Do you find any interesting relationships? Did you know that every Pythagorean triple has at least one element that is divisible by either 3, 4, or 5? For example, (12, 35, 37) has 12, which is divisible by both 3 and 4, and 35, which is divisible by 5. Notice that the product of all three numbers is always a multiple of 60.

11.6 RECIPROCAL TRIPLES

Pythagorean triples are triples (A, B, C) of positive integers so that

$$A^2 + B^2 = C^2$$

Examples of Pythagorean triples are (5, 12, 13) and (6, 8, 10), since $5^2 + 12^2 = 13^2$ and $6^2 + 8^2 = 10^2$. The triple with the smallest sum is (3, 4, 5).

Interestingly, reciprocal Pythagorean triples exist. These are triples (A, B, C) of positive integers with the property

$$\frac{1}{A^2} + \frac{1}{B^2} = \frac{1}{C^2}$$

An example of a reciprocal Pythagorean triple is (156, 65, 60) since

$$\frac{1}{156^2} + \frac{1}{65^2} = \frac{1}{60^2}$$

The following program finds two reciprocal Pythagorean triples with sums less than 100. The program will find more triples if you increase the limit in the first **for** statement.

```
program ReciprocalTriples;
var
  A      :Integer;(* Triple A *)
  B      :Integer;(* Triple B *)
  C      :Integer;(* Triple C *)
  X      :Real;(* A*A *)
  Y      :Real;(* B*B *)
  Z      :Real;(* (A*A * B*B) / (A*A + B*B) *)
begin
  writeln('A          B          C          A + B + C');
  writeln('------------------------------------------');
  for A := 1 to 40 do
    for B := 1 to A do
      begin
        X := A * A;
        Y := B * B;
        Z := (X * Y) / (X + Y);
```

```
        C := TRUNC(SQRT(Z));
        if C * C = Z then
          if A + B + C < 100 then
            writeln(A,B:7,C:9,A + B + C:12);
      end;
end.
```

COMPILE/RUN

A	B	C	A + B + C
20	15	12	47
40	30	24	94

Eyn Newe
Unnd wolgegründte
underweysung aller Kauffmanß Rech-
nung in dreyen Büchern/mit schönen Re-
geln vñ fragstucken begriffen. Sunder-
lich was fortl unnd behendigkait in der
Welschē Practica vñ Tolleten gebraucht
würde / des gleychen fürmalß weder in
Teützscher noch in Welscher sprach nie
gedrückt. durch Petrum Apianū
von Leyßnick/d Astronomei
zu Ingolstat Ordina-
riū / verfertiget.

Pascal's triangle first appeared in print on the title page of the *Arithmetic of Petrus Apianus*, Ingolstadt, 1527, more than a century before Pascal investigated the properties of the triangle.

Blaise Pascal was born June 19, 1623, in Clermont-Ferrand, France. While the other youngsters were amusing themselves at play, Pascal was off in a corner working out complicated mathematical problems. While still a child, he created many theorems that were identical to those found in the first book of Euclid. Pascal, who never attended school, went on to become one of France's greatest mathematicians and philosophers. Pascal is probably best known for his discovery of a new branch of mathematics which came to be known as *probability*.

11.7 PASCAL'S TRIANGLE

Determining the chance that something will happen is like looking into the future. It is done by using common sense and a knowledge of what happened in the past. To see how it works in a simple case, let us try to foresee what happens when you toss a coin. The coin has two faces, head and tail. Common sense and experience join to tell us that, out of a large number of tosses, about half will come out heads, and the rest will be tails. Saying it another way: on the average, one out of two tosses will come out heads. So we say the chance of getting a head is 1/2.

If we toss two coins, there are three possible results. We may get two heads, or two tails, or one head and one tail. The chance of getting two heads is one out of four, or 1/4. The chance of getting two tails is also 1/4. The chance of getting one head and one tail is two out of four, or 1/2. What is the chance of getting two heads and a tail when you toss three coins? If you toss four coins, what is the chance of getting all heads or all tails?

There is a short method for finding the answers to these questions in the arrangement of numbers known as Pascal's triangle. Pascal, a French philosopher and mathematician of the seventeenth century, was for a time interested in games of chance. This interest led him to discover certain important rules about the probabilities of getting heads or tails on the toss of a coin. His findings are described in the triangle formation of numbers on page 217. The formation is easy to construct and shows the chance of getting heads or tails, or any combination of them, on a given number of tosses of a coin.

If you toss one coin, the chance of getting heads is 1 out of 2, or 1/2. If you toss two coins, your chance of getting 2 heads is 1 out of 4; of getting 1 head and 1 tail, 2 out of 4 or 1/2; of getting 2 tails, 1 out of 4. If you toss three coins, your chances are: all heads, 1 out

Pascal's triangle — a pattern of numbers from which binomial probabilities can be easily determined. It was known to Omar Khayyam about 1100 A.D. and was published in China about 1300 A.D. However, it is generally known as Pascal's triangle because of the amount of work he did on it.

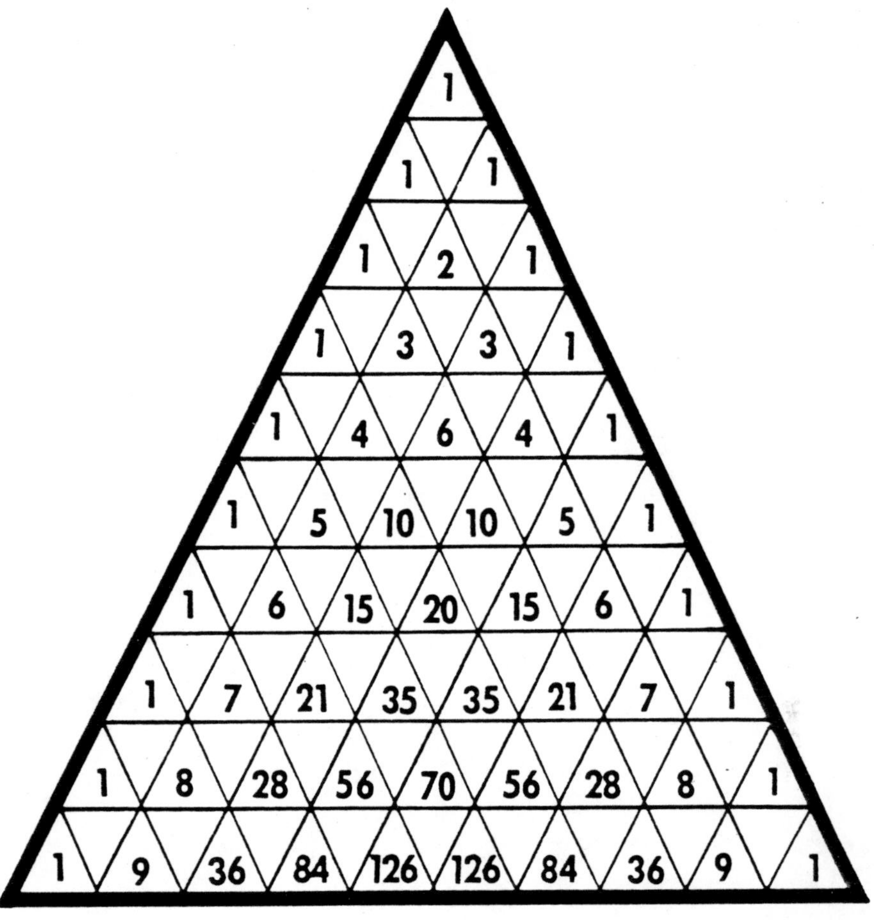

	1	2	3	4	5	6	7	8	9	10
1	1	1	1	1	1	1	1	1	1	1
2	1	2	3	4	5	6	7	8	9	
3	1	3	6	10	15	21	28	36		
4	1	4	10	20	35	56	84			
5	1	5	15	35	70	126				
6	1	6	21	56	126					
7	1	7	28	84						
8	1	8	36							
9	1	9								
10	1									

Pascal constructed the triangle as in the above figure, each horizontal line being formed from the one above it by making every number in it equal to the sum of those above and to the left of it in the row immediately above it; for example, the fourth number in the fourth line, namely, 20, is equal to 1 + 3 + 6 + 10. The numbers in each line are **figurate** numbers. Those in the first line are called numbers of the first order; those in the second line, natural numbers or numbers of the second order; those in the third line, numbers of the third order, and so on. The *m*th number in the *n*th row is $(m + n - 2)! / (m - 1)! (n - 1)!$.

of 8; 2 heads and 1 tail, 3 out of 8; 2 tails and 1 head, 3 out of 8; all tails, 1 out of 8.

If four coins are tossed, there is 1 chance in 16 of getting all heads or all tails; 4 out of 16 of getting 3 heads and 1 tail, or 3 tails and 1 head; and 6 out of 16 of getting 2 heads and 2 tails. In five tosses, chances are: 1 out of 32 for all heads or all tails; 5 out of 32 for 4 heads and 1 tail, or 4 tails and 1 head; and 10 out of 32 for 3 heads and 2 tails or 3 tails and 2 heads.

There are many different ways to generate the triangle: combinatorial methods, trigonometric methods, or the binomial theorem method. A very easy way of constructing the triangle exists. Each number within the triangle is found by adding the two numbers above it at the left and right. The coefficients in the expansion of $(A + B)^N$ as N is assigned the values 0, 1, 2, . . . successively, can be used to form the triangular array. The number of combinations of N things taken R at a time (a basic problem in probability theory) can also be used to produce the coefficients of this expansion.

The following program produces the numbers of Pascal's triangle.

NUMBER THEORY TRIVIA

If you had a large Pascal's triangle of binomial coefficients, you might notice that all of the entries in rows 1, 3, 7, 15, 31, 63, . . . (these numbers of the form $2^n - 1$ are called Mersenne numbers) are odd integers and each of the other rows had at least one even entry.

```
program PascalsTriangle;
var
  Triangle           :array[1..13, 1..13] of Integer;
  I,J,K,L,Spaces     :Integer;
begin
  for I := 1 to 13 do
    begin
      Triangle[1,I] := 0;
      Triangle[I,1] := 1;
    end;
  for J := 2 to 13 do
    for K := 2 to 13 do
      Triangle[J,K] := Triangle[J-1,K-1] + Triangle[J-1,K];
  Spaces := 30;
  for J := 1 to 13 do
    begin
      for L := 1 to Spaces do
        write(' ');
      Spaces := Spaces - 2;
      for K := 1 to J do
        write(Triangle[J,K]:4);
      writeln;
    end;
end.
```

COMPILE/RUN

```
                              1
                            1   1
                          1   2   1
                        1   3   3   1
                      1   4   6   4   1
                    1   5  10  10   5   1
                  1   6  15  20  15   6   1
                1   7  21  35  35  21   7   1
              1   8  28  56  70  56  28   8   1
            1   9  36  84 126 126  84  36   9   1
          1  10  45 120 210 252 210 120  45  10   1
        1  11  55 165 330 462 462 330 165  55  11   1
      1  12  66 220 495 792 924 792 495 220  66  12   1
```

Pascal showed an early interest in geometry, and at the age of sixteen wrote an essay containing his famous hexagon theorem. Shortly afterwards he invented a calculating machine. His many and varied contributions include the study of atmospheric pressure, integration, the theory of probability, and the properties of curves such as the cycloid. He is also credited with inventing the syringe, hydraulic press, and the wheelbarrow. This postage stamp was issued by France.

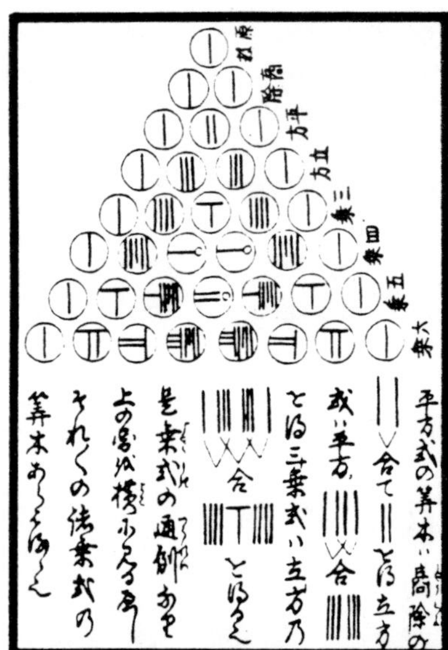

The Japanese version of the triangle from Murai Chuzen's *Sampo Doshi-mon* (1781). This illustration also shows the Sangi board (counting board ruled in columns) forms of the numerals.

11.8 CROSSING THE RIVER

The following drawing shows a computer graphic solution to a well-known riddle.

NUMBER THEORY TRIVIA

The smallest palindromic prime number that is composed of all four prime digits (2, 3, 5, 7) is

7352537

Can you write a program to produce the next palindromic prime that contains all four prime digits?

Pierre de Fermat (1601-1665) was a lawyer by profession, but he was an amateur mathematician in his spare time. He became Europe's finest mathematician, and he wrote well over 3000 mathematical papers and notes. However, he published only one, because he did them just for fun. His scientific and mathematical achievements were many and varied, but his great love was number theory.

A farmer owns a dog, a duck, and a bag of corn, and he wishes to transport them to the other side of a river. Unfortunately, the only available boat is large enough to hold only the farmer and one of his possessions at one time. (It is a large bag of corn!) The farmer cannot afford to leave the dog and the duck together unchaperoned, for the former would eat the latter. Similarly, the duck and bag of corn may also not be left together unattended. How can the farmer safely transport all three of his possessions across the river?

You may wish to program a computer to provide a numerical solution to this riddle.

11.9 FERMAT'S AND GOLDBACH'S CONJECTURES

Probably no branch of mathematics other than number theory is so replete with unsolved problems that are so simple as to be readily understood by someone without advanced training. Some of the unsolved problems are so significant and deep that a solution would be a major world event (well, at least in math circles). Others remain unsolved or even forgotten mainly because mathematicians consider them too dull or trivial to justify working on them. Still others lie within the grasp of even those of us who aren't math prodigies. Two of the more famous unsolved problems are described in this section.

Fermat's Last Theorem

Consider the equation: $a^n + b^n = c^n$. If $n = 2$, it has an infinity of solutions in integers. The simplest is $3^2 + 4^2 = 5^2$. Such solutions are called Pythagorean triples because they measure the sides of right triangles. Are there solutions in integers if n is greater than 2? The famous seventeenth century French mathematician, Pierre Fermat, scribbled in the margin of a book that he had discovered a "truly marvelous" proof that the answer is no, adding that the margin was too small to contain it. Fermat never disclosed his proof.

To this day, no one has been able to prove Fermat's last theorem, as it came to be known, or to find a solution that falsifies it. Computers have made exhaustive searches up to exponents of 125,000 and values for a, b, and c that are millions of digits long.

Fermat's conjecture is probably true, but it could be what mathematicians call an "undecidable" true theorem. If so, the prospects for proving it are bleak. For centuries mathematicians will struggle vainly to construct a proof, and of course they will never find a counterexample. If the theorem is false, of course, it cannot be undecidable, because a single counterexample would decide it. In 1983, a young West German mathematician, Gerg Faltings, made some progress by proving that if the theorem is false, the equation has only a finite number of basically different solutions for each exponent.

Goldbach's Conjecture

Christian Goldbach (1690-1764), a Russian mathematician, made a shrewd guess - in mathematics it is called a conjecture - that every even number is the sum of two prime numbers. For instance, 12 = 5 + 7 and 18 = 5 + 13.

Goldbach communicated his guess to his illustrious friend Leonhard Euler (1707-1783), a Swiss mathematician, who was quite impressed. The surmise seemed to him to be a true proposition, but his sustained efforts to prove it and the efforts of his followers up to the present time have been in vain.

In the late 1930s, I. M. Vinogradov proved that any odd number is the sum of three prime numbers. Thus 17 = 3 + 7 + 7, 19 = 3 + 5 + 11, 21 = 3 + 7 + 11, and 35 = 5 + 7 + 23. The complete problem is still unsolved. Perhaps you would like to write a program to show that several hundred even numbers greater than 2 are the sum of two primes.

The following program prints one sum of prime numbers for each even number from 6 through 100.

```
program GoldbachsConjecture;
var
  P                    :array[1 . . 50] of Integer;
  C,I,J,K,N,X,Y,T      :Integer;
begin
  C := 0;
  writeln('EVEN NUMBERS WITH ASSOCIATED PRIME NUMBERS');
  writeln('------------------------------------------------------');
  (**** Store prime numbers in Array P ****)
  K := 1;
  while K < 50 do
    begin
      N := 2 * K + 1;
      J := 1;
      I := 3;
      while(I*I <= N) and (J = 1) do
        begin
          if N mod I = 0
            then J := 0 else I := I + 2;
        end;
      if J = 1 then
        begin;
          C : = C + 1;
          P[C] := N;
        end;
      K := K + 1;
    end;
  (**** Find one sum of odd primes for each even number ****)
  T := 6;
```

NUMBER THEORY TRIVIA

There are many composite palindromic integers which can be factored into palindromic primes. For example:

1111 = 11 × 101
1441 = 11 × 131
3443 = 11 × 313

Can you write a program to find other composite palindromic integers whose palindromic prime factors consist of digits greater than 3?

```
  repeat
    for X := 1 to C do
      for Y := X to C do
        if T = P[Y] + P[Y] then
          writeln(T:2,' = ',P[X]:2,' + ',P[Y]:2);
    T := T + 2;
  until T > 50;
end.
```

11.10 PALINDROME CONJECTURE

A palindrome is a word, verse, or number that reads the same backwards or forwards. For example, the words eye, dad, tot, radar, toot, and mom are palindromes. So are each of the following arrangement of words:

MADAM, I'M ADAM
STEP ON NO PETS
DON'T NOD
WAS IT A RAT I SAW
NEVER ODD OR EVEN
LIVE NOT ON EVIL
TOO HOT TO HOOT

Numeric palindromes are those numbers which read the same backward as forward. The examination of these numbers is a field rich with possibilities for creative problem solving with computers. A numeric palindrome is a positive integer with at least two digits. Some examples of numeric palindromes are these:

33
171
555
3884
93539
1234321

A mathematical conjecture describes how a palindrome can be obtained from a non-palindromic number: begin with any positive integer that contains at least two digits. If it is not a palindrome, reverse its digits and add the two numbers. If the sum is not a palindrome, treat it as the original number and continue. The process stops when a palindrome is obtained. For example, beginning with 87:

Not a palindrome	87
	78
Not a palindrome	165
	561
Not a palindrome	726

NUMBER THEORY TRIVIA

A number is a **palindrome** if it is unchanged by reversal, i.e., 121, 14541, etc. Palindromic numbers which are also powers, for example

$$(11)^2 = 121$$

have several interesting number theoretic properties. The following table shows numbers where both the numbers and the powers are palindromic.

x	x^2
1	1
2	4
3	9
11	121
22	484
26	676
101	10201
111	12321
121	14641
202	40804
212	44944
264	69696
307	94249
836	698896
1001	1002001
1111	1234321
2002	4008004
2285	5221225
2636	6948496
10001	100020001
10101	102030201
10201	104060401
11011	121242121
11111	123454321
11211	125686521
20002	400080004
20102	404090404
22865	522808225
24846	617323716
30693	942060249
100001	10000200001
101101	10221412201
110011	12102420121
111111	12345654321
200002	40000800004
798644	637832238736
1000001	1000002000001
1001001	1002003002001
1002001	1004006004001
1010101	1020304030201
1011101	1022325232201
1012101	1024348434201
1042151	1086078706801
1100011	1210024200121
1101011	1212225222121
1102011	1214428244121
1109111	1230127210321
1110111	1232346432321
1111111	1234567654321
1270869	1615108015161
2000002	4000008000004

	627	
Not a palindrome	1353	
	3531	
	4884	Is a palindrome

The conjecture, often assumed true, is that this process will always lead to a palindrome. And indeed that is just what usually happens. Most numbers less than 10,000 will produce a palindrome in less than 24 additions.

You will notice that when there are no carries in the addition, as in 1353 + 3531, it is automatic that the result is a palindrome.

The number 89 requires twenty-four reversals and sums before the palindrome is obtained.

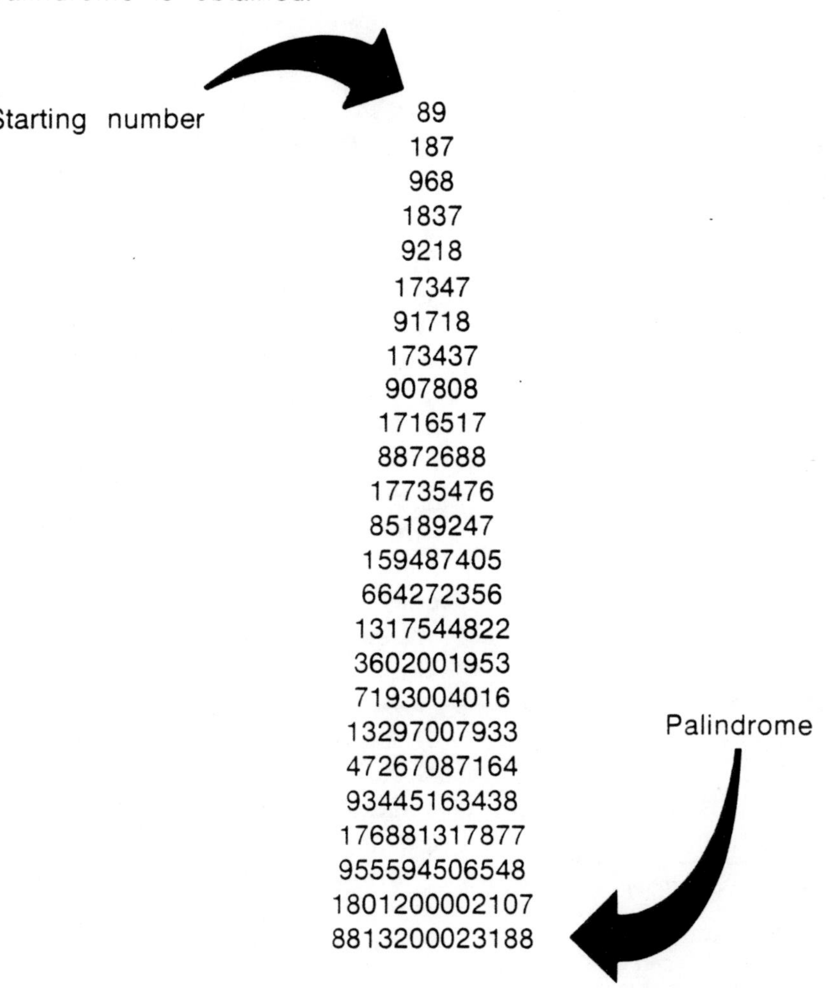

For decades it was assumed that every starting number produced a palindrome. The mathematician; Charles Trigg, wasn't so sure. He began testing numbers in consecutive order and found that all of them reached a palindrome in six or fewer steps until he came to 89. As was shown above, 89 requires twenty-four steps to produce

NUMBER THEORY TRIVIA

There are ten palindromic numbers (6 digits or less) whose cubes are also palindromes. Three of these numbers are shown below.

$11^3 = 1331$
$111^3 = 1367631$
$10001^3 = 1000300030001$

Can you write a program to find the other seven palindromic numbers?

a palindrome. Trigg found that its reversal, 98, also took 24 steps. From there on, palindromes appeared in no more than 24 steps until 196. To Trigg's amazement, after 100 steps there was still no palindrome. He found 149 integers under 1000 that seemed unable to generate a palindrome. By 1967 he was convinced that the palindrome theorem was false.

In 1975 Harry Saal, at IBM's Israel Scientific Center, used a computer to test 196 to 237,310 steps, reaching a final sum with 98,305 digits. Still no palindrome. No one had yet proved the existence of a number that would never reach a palindrome, although curiously, such numbers have been shown to exist in any system of notation based on a power of 2. For instance, if you start with the binary number 10110 (or 22 in decimal notation), the sums fall into a four-step cycle that keeps extending the basic pattern that isn't reversible. The question remains unanswered for all other notation systems.

If you would like to make a number theory breakthrough, you might find the 196 problem or other palindromic number challenges interesting programming challenges.

11.11 HARD PROBLEMS

Computers have expanded mathematicians' horizons, allowing them to make calculations never before dreamed of. But, at the same time, they have made mathematicians recognize the limits of their ability to solve certain types of problems. Within the past decade, mathematicians and computer scientists grouped together hundreds of related problems. These, in principle, can be solved primarily by adding and multiplying. However, even the best methods of solving these problems can require billions upon billions of calculations - enough to keep the computers busy for years, even centuries. The scientists are now learning to live with this impediment. And perhaps they can even exploit it to create a new kind of seemingly unbreakable secret code.

These simple but possibly unsolvable problems are not new. Many have been around for decades. They crop up in many practical situations. But until 1971, mathematicians and computer scientists did not realize that the problems were related. Then, Stephen Cook of the University of Toronto made a discovery. He found that several of these problems were equivalent. This means that if anyone could find a shortcut to solving one of them, the shortcut could be adapted to solve the others.

NUMBER THEORY TRIVIA

What is the minimum number of bins, each having the capacity 50,000,000,000, such that a numerical value can be placed into them? It has been stated that "even with the use of all the computing power in the world currently available, there seems to be little hope of knowing the answer to this question." This task will definitely be a challenge to future mathematicians and supercomputers.

Previously mathematicians had been looking at each problem separately, hoping somehow to find a way to solve it in a feasible length of computer time. But Cook's discovery systematized the study of these problems. Shortly afterwards, Richard Karp of the University of California at Berkeley greatly extended the list of equivalent problems. Then, in the scientific community, a scramble to find which problems were equivalent to these hard ones began. So far, hundreds have been added and more are under consideration.

Here is a classic hard problem. Suppose a salesperson wants to make exactly one visit to each capital in the 48 contiguous states, starting and ending at the same capital. What is the shortest route that would work? The problem remains unsolved because of the amount of calculation needed — the salesperson must consider 24!/3 possible routes, a number with 60 digits.

Hard problems are technically called NP-complete (NP means nondeterministic polynomial) by mathematicians and computer scientists. What sort of problems are classified hard?

One example is the traveling salesman problem: a salesman wants to plan a tour of a number of cities so that he visits each city only once and wants to find the shortest possible route. This problem turns up in numerous guises in practical situations. The telephone company must solve a traveling salesman problem when it plans collections from pay telephone booths. The telephone company divides each city into zones. Each zone contains several hundred coin boxes. The company supervisors must decide the best order to collect coins from the telephones in each zone.

Another hard problem is the bin packing problem: suppose there are a given number of identical bins and a group of odd-shaped packages. What is the minimum number of bins necessary so that each package is in a bin, and none of the bins overflow? It is a bin packing problem to decide how to schedule television commercials to fit in one minute time slots. It is also a bin packing problem to find out how to cut up the minimum number of standard length boards to produce pieces of particular lengths.

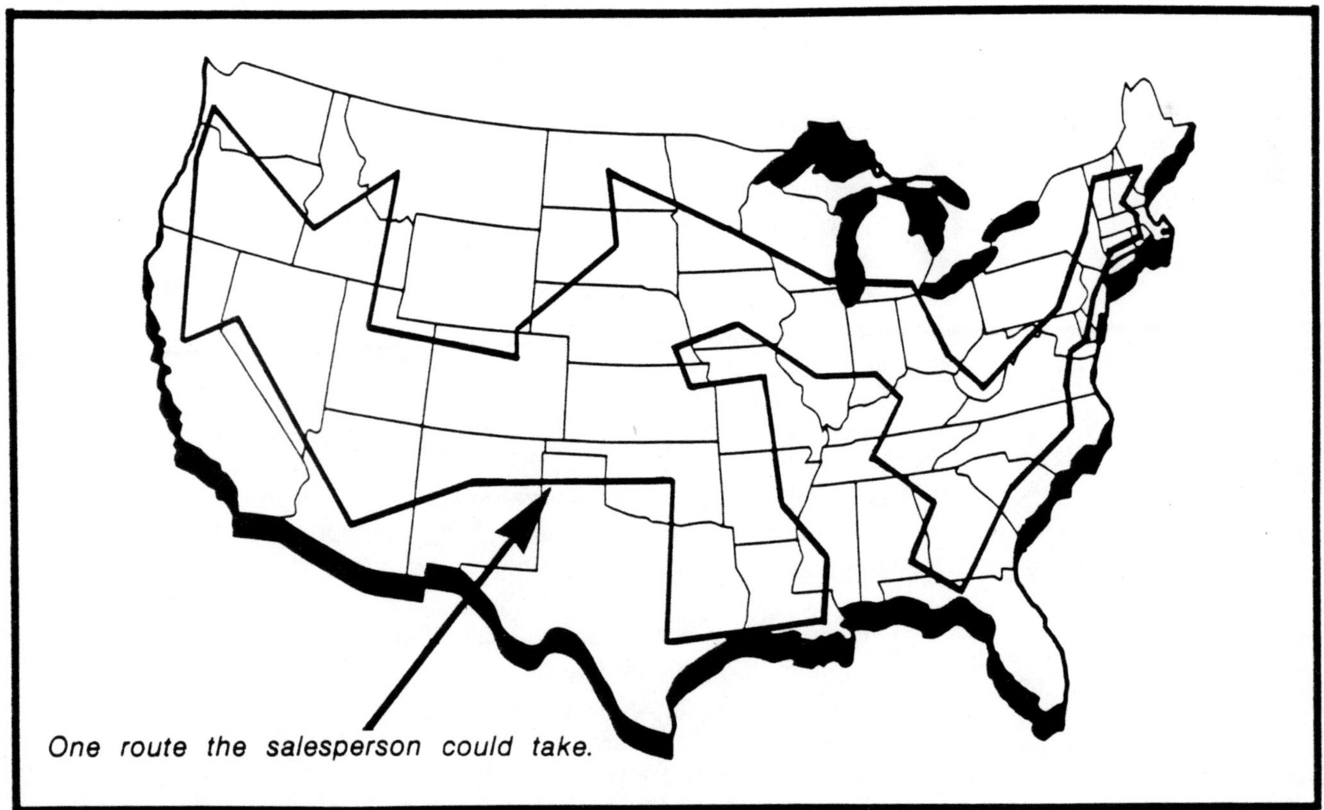

One route the salesperson could take.

NUMBER THEORY TRIVIA

There is a set of palindromic prime numbers that are composed of prime digits. For example,

2
3
5
7
353
373
727
757
32323
33533
35353
35753
37273
37575

Can you write a program to produce additional palindromic primes composed of prime digits?

The only general method of solving these hard problems is to try all possible solutions until you find the best one. For small problems, this isn't too hard. If a salesman had to visit only four cities, he could plan his tour himself with just a pencil and paper by considering all 24 possibilities.

But if the salesman had to visit ten cities, it would be considerably harder to try out all routes, because there would be more than 3,500,000 of them. The task is, however, well within the capabilities of a computer. But if a salesman had to visit 18 cities, and he had a computer that could test one million routes per second, it would take the computer about 4000 years to try all possible routes. If the salesman had to visit 60 cities, it would take a computer billions of centuries to try out all the routes.

Why does it take so long to try out all possible solutions to a hard problem? The reason is that the number of possible solutions increases explosively as the size of the problem grows. In a bin packing problem, for example, this rate of increase is found by multiplying a certain fixed number by itself each time you add another object to be packed. The fixed number equals the number of bins. The number of multiplication steps required varies with the number of objects. Thus, if you have two bins and two objects, the number of possible solutions is four (2 × 2); with three objects, there are eight possible solutions (2 × 2 × 2). Each object you add multiplies the number of possible solutions.

In a bin packing problem of this kind, computer scientists have calculated that with ten objects, the possible solutions could be tried in 1/1000 of a second. With 20 objects, trying all the solutions would take about one second, with 30 objects, about 18 minutes would be needed, and with 100 objects to pack, trying all the possible solutions would take more than 400,000 billion centuries. (By contrast, the entire universe is only about 10 billion years old.)

Because the hard problems are of practical importance, many computer scientists and mathematicians are spending considerable amounts of time trying to develop methods that will solve some of them. Obviously, exponential time methods are not practical for large problems. What mathematicians would like to find is a polynomial time method that would require much less computer time. Computer scientists calculate that a problem that would take billions of centuries to solve by an exponential method might be solved in seconds if a polynomial method were found.

An exciting spin-off from the discovery of hard problems is the idea of using them as the basis of a new kind of secret code. Such a code was first suggested a few years ago by Whitfield Diffie, Martin Hellman, and Ralph Merkle of Stanford University. The reason the code may be unbreakable is that an eavesdropper trying to decipher a message would have to solve a hard problem to do so.

REVIEW EXERCISES

1. What is a Diophantine equation?
2. Find a solution in integers of $5x + 7y = 29$.
3. Find a solution in integers of $33x + 14y = 173$.
4. Write a program that will find Pythagorean triangles whose sides are in arithmetic progression. Hint The sides are any numbers proportional to 3, 4, 5.
5. Write a program that will find primitive Pythagorean triangles whose perimeter is a square. Hint the sides are $b^2(4a^2 - b^2)$, $4a^2(b^2 - 2a^2)$, and $8a^4 - 4a^2b^2 + b^4$ where b is odd and less than 2a but greater than a times the square root of 2. For example, if $a = 2$ and $b = 3$ then the sides of the triangle are 63, 16, and 65.
6. Write a program that will find the four smallest primitive Pythagorean triangles that have the same perimeter.
7. Briefly describe Pascal's triangle.
8. Modify the Pascal's Triangle program so that it will produce a triangle with ten rows.
9. What is Fermat's last theorem?
10. Write a program to show that the even numbers between 6 and 300 are the sum of two primes.
11. What is a palindrome?
12. Is the number 78 a palindrome?
13. Show that the number 19 will produce a palindrome.
14. How many steps are required to obtain a palindrome from the number 68?
15. The number 89 will produce the palindrome 8813200023188 after ____________ steps.
16. Show that the number 79 will produce a palindrome after 6 steps.
17. What is meant by a "hard problem?"

NUMBER THEORY TRIVIA

Take an integer, reverse the digits, and add the reverse number to the original. This process is continued until the result is a **palindromic number**. For example,

```
 39
+93
----
132
+231
----
363
```

Can you write a program to find the smallest two consecutive starting numbers that produce the same palindrome?

HAVE YOU EVER SOLVED A HARD PROBLEM?

ALL PROBLEMS ARE **HARD**.

12 AN INTRODUCTION TO PASCAL

PREVIEW

Pascal is a general-purpose programming language with a small number of simple but powerful features. Pascal was the first major new language to be developed after the concept of structured programming was introduced. The language was named after the French mathematician and philosopher Blaise Pascal, who, among other things, invented a calculating device.

Pascal was originally developed as a teaching language for students of computer science. It has the elements of simplicity, and, at the same time, provides a wide range of data types and structures which can be used to very good effect. Until the coming of relatively inexpensive microcomputers, the use of computers was restricted to those people fortunate enough to be working for or studying in an organization that owned one of these expensive machines, and was prepared to make a small amount of time available for programs to be written.

With the coming of the microcomputer, all this has changed. It is now possible to purchase low-cost machines that will perform extremely complex tasks and store large volumes of information. The problem is that once the machine is available, it must be told what it is expected to do. This requires the writing of a program. While most of the tasks that a microcomputer will perform will be adequately managed by the purchase of commercial packaged programs (such as word processing programs, spreadsheet programs, data base management programs, drawing programs, etc.), the full potential of the microcomputer can only be realized if some programs can be written for it. The great advantage of Pascal is that it is possible to provide a full compiler, together with the filing system and text editor that makes a complete programming environment, all within the constraints of the currently available microcomputers.

Blaise Pascal (1623-1662) made revolutionary contributions to mathematical thinking. In 1641, when he was only eighteen, he invented an arithmetic machine. His gadget made of metal cogs and wheels could add and subtract rapidly, with perfect accuracy.

The Pascal programming language was originally specified in 1968 by Professor Niklaus Wirth of Switzerland. The language was designed to satisfy two principal goals: (a) institute a language suitable for teaching programming as a systematic discipline; and (b) define a language whose implementations would be reliable and efficient on computers that were currently available. The success in achieving these goals can be measured by the rapid and widespread increase in the use of Pascal as a practical language for writing systems and applications programs, and for teaching the principles of computer programming. The first Pascal compiler became available in 1970. By 1971, the language was formally defined. Over the next decade, continual changes made the language less machine-dependent. Today, Pascal exists for all hardware from microcomputers to supercomputers, and its range of uses has broadened considerably from its original role in teaching and systems programming. Current uses of Pascal range from business systems to real-time processing for NASA's space programs.

After you finish this chapter, you should be able to:

1. Understand the basic structure of a Pascal program.
2. Understand the most common statements in the Pascal language.
3. See how Pascal programming can be used to solve number theory problems.
4. Write Pascal programs to solve simple problems in number theory and mathematics.
5. Interpret simple programs written in the Pascal language.

12.1 GETTING STARTED IN PASCAL

Pascal Features

Pascal features a number of general characteristics which are desirable in a programming language:

- It has relatively few concepts and has simple and regular rules for combining them. A programmer can readily use these concepts as primitives in developing algorithms.
- Pascal enables programmers to write readable programs following good program design techniques. Algorithms can be directly translated into Pascal and the program structure will reflect the underlying logical structure of the algorithm.
- The two previous characteristics simplify the process of program testing. Verifying that a program, or a portion, is correct (that it does what it is intended to do) not only can guarantee good results but also can reduce costly program maintenance when the program is used over a period of time.
- Programs written in Pascal can be run on many different computers. This transportability is a very desirable feature.

Niklaus Wirth

In 1968, after five years as a visiting professor at Stanford University, Swiss computer scientist Niklaus Wirth returned to Switzerland to begin work on Pascal. The language would soon become a standard tool for teaching computer programming. In 1981, he announced Modula-2, a language he had forged to replace Pascal for general purpose applications. Modula-2 expanded upon Pascal in a number of ways, most notably by providing for large and complex programs that could be composed in separate, self-contained modules and then fitted together.

Basic Data Types In Pascal

A Pascal program can process many different kinds of data. The data are represented differently inside the computer and manipulated using different operations. To help us deal with the diversity of data and avoid mistakes such as trying to add or multiply letters of the alphabet, Pascal classifies data into different data types. Items belonging to the same data type can be manipulated using the same operations and can be represented in similar ways both in Pascal programs and inside the computer.

Pascal has four standard data types: integer, real, Boolean, and character. Integer, Boolean, and character are called **scalar values**. Scalar data are values that imply counting, and have a known range and number of possible values.

Philippe Kahn

A former student of Niklaus Wirth, developed a Turbo Pascal compiler for microcomputers. It was compact, lightning fast and remarkably inexpensive. Turbo Pascal is the most popular version of the Pascal language for microcomputers.

Integer Data Types **Integers** are whole numbers that do not contain decimal points. However, an integer constant may be preceded by a + or − sign. When the sign is omitted, a plus sign is assumed. Thus 36 and +36 represent the same value. The following are examples of integer constants:

1048 26 4639 −842 +6149

Commas are not allowed in writing integers. Thus 21,018 would have to be written as the integer constant 21018.

Real Number Data Types **Real constants** are numbers that can have fractional parts. The following are examples of real constants:

24.9 300.0 −28.34 0.05 +52.5

Real constants always include a decimal point and many contain an exponent (a number represented by a base, mantissa, and a character or exponent). For example, the value 86.3×10^2, where 86.3 is the mantissa, 10 is the base, and 2 is the exponent, would be represented in Pascal by the real constant 86.3E2, where 86.3 is the mantissa, 2 is the exponent, and E means *with exponent*. The number is to the base 10. This method of expressing real constants is known as **floating-point notation** or scientific notation.

There are several ways of representing the same value. For example, the values −1.5E3, −15.0E2, −150.0E1, −1500.0E0 are all equivalent to the value −1500. In a real value, the mantissa and the exponent may be either positive or negative. This form of constant allows us to express very large or very small numbers. Here are a few more examples of floating-point notation:

Conventional Notation	Floating-Point Notation
700.0	7.0E2
582.23	5.8223E2
0.038	3.8E−2
934000000	9.34E8

Boolean Data Types An important feature of computers is their ability to take different actions depending on the conditions that hold true when a program is executed. This ability allows us to program the computer to make decisions based on input values. The logical capabilities of Pascal allow one to perform operations that deal with logical relations between pieces of information. The result of such an operation is an indication that the relationship was either *True* or *False*. The constants True and False are often called truth or logical values, but in Pascal they are called **Boolean values.**

Character Data Types Most computer applications involve alphabetic information as well as numeric information. A program designed to maintain mailing lists, for example, will require a capability for accepting alphabetic characters as input data, storing them, and printing them out again. Before we can write programs that manipulate characters, we will need a data type to represent them. For this purpose, Pascal provides the data type **char**. Rather than holding a numeric or truth value, a character data type will

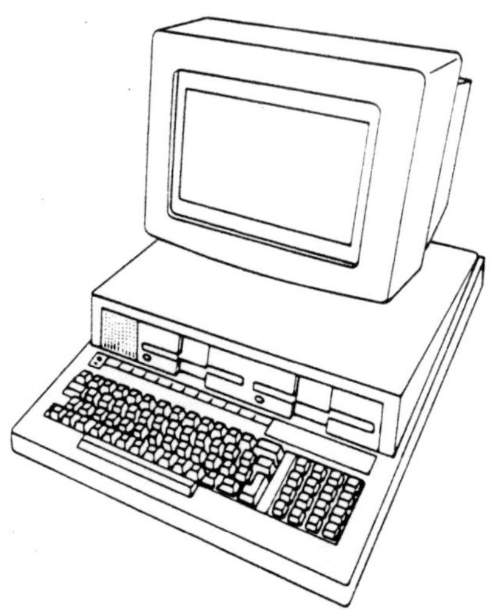

Most microcomputers support a variety of programming languages, but BASIC and Pascal continue to be the preferred languages with micro users.

hold a character, such as a letter, a digit, or a symbol. For example, while the data type integer can hold the value 6, the data type char can hold the character "6." And while numeric values can be added together, multiplied, divided, etc., characters can be combined into strings of characters, compared to each other, and manipulated in other ways. When the numeric data type 6 is added to 6 it yields the value 12. When the character "6" is added to the character "6" the result is the string "66."

When using character data types, the characters themselves must be placed in quotes.

'7' '+' 'R' 'h'

The single quotation marks are necessary so that the Pascal compiler will know, for example, that 7 is a character constant and not the integer seven, or that + is a character constant and not an addition sign.

A series of characters, such as a word or sentence, is called a **string.** A string constant consists of a series of characters enclosed in single quotation marks. The following are examples of string constants:

'WHAT IS YOUR NAME?'
'BETTY WINSLOW'
'PRIME NUMBERS'

Identifiers

In Pascal programs many programs, constants, memory locations, variables, data types, procedures, and functions have names. These names are called **identifiers.** Identifiers must be formed according to certain rules, so the Pascal compiler can distinguish them from other parts of a program. In Pascal an identifier is made up of a letter followed by up to 126 characters (letters, numbers, and underscores). The case of letters is irrelevant; therefore, the identifiers BLACKJACK, BlackJack, Blackjack, and blackjack are all the same. Some legal identifiers are

Sum
R635
volumn
Message
FibonacciNumber
StandardDeviation
Grade
root

The following identifiers are invalid:

hot-stuff (Hyphen not allowed)
Fibonacci Numbers (Space is not permitted)
total$ ($ not allowed)
8R (Must begin with a letter)

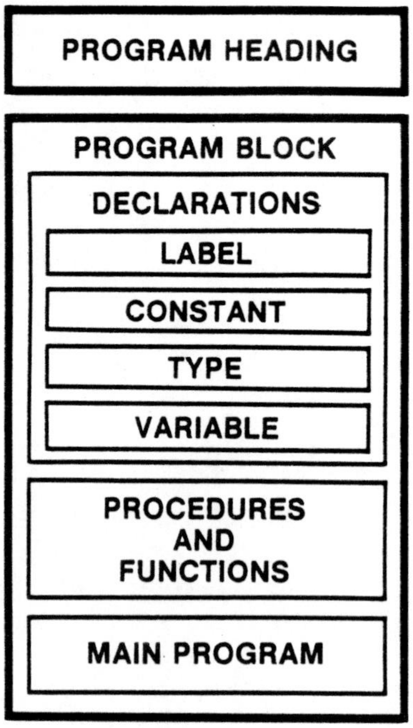

PROGRAM HEADING

PROGRAM BLOCK
DECLARATIONS
LABEL
CONSTANT
TYPE
VARIABLE
PROCEDURES AND FUNCTIONS
MAIN PROGRAM

The overall organization of a Pascal program.

Parts Of A Pascal Program

A Pascal program consists of a program heading followed by a program block. The program block is further divided into a: (a) declaration part, in which all objects local to the program are defined; and (b) statement part, which specifies the actions to be executed upon these objects. The overall shape of a Pascal program is illustrated in the margin.

The Program Header Each Pascal program should begin with a program header. The program header, which is always the first line in a program, contains the word **program** followed by the name of the program and a semicolon. Although optional, the program header is recommended because it gives the program an identity. The semicolon (;) separates the program statement from the rest of the program. Some examples of program headings are

```
program Polynomial;
program PythagoreanTriples;
program PalindromicNumbers;
program LeastCommonMultiple;
```

Declarations Pascal requires that every item of data and every program module be declared before it is used in the execution of the program. Data declaration simply means informing the Pascal compiler what data it will have to accept while processing the program so that the computer can allocate the correct amount of space in computer memory to accomplish the task as efficiently as possible. Because some kinds of data take up more space than others, the compiler needs to know not only what data it will be seeing, but what kind of data it is.

The declaration part of a Pascal program is divided into four parts or sections:

- Label declaration
- Constant declaration
- Type declaration
- Variable declaration

The **label declaration** section of a program consists of the word **label** followed by one or more label identifiers, separated by commas, and terminated with a semicolon. The purpose of a label is to mark the destination of a **goto statement**.

The **constant declaration** section introduces identifiers as synonyms for constant values. The term **const** heads the constant declaration section and is followed by a list of constant assignments separated by semicolons. Each constant assignment consists of an identifier followed by an equal sign and a constant. Here are some examples:

```
const
  Limit = 1000;
  Author = 'Nancy Dugan';
```

As shown here, constants can be either numbers or strings.

NUMBER THEORY TRIVIA

What's in a name? Have you ever noticed that when you hear the last names of certain mathematicians, their first names spring to mind? (Fermat . . . Pierre, Euler . . . Leonhard). Yet other eminent mathematicians whose last names slip off our tongues every day are not so familiar. The following list contains the last names of "well known" mathematicians. Can you supply the first names? They are all in this book.

Gauss
Pascal
Slowinski
Fibonacci
Goldbach
Mersenne
Wirth
Viete
Leibniz
Wallis
Lagrange

NUMBER THEORY TRIVIA

The following two numbers can be expressed as the sum of two squares:

$$5882353 = 588 \times 10^4 = 588^2 + 2353^2$$

and

$$99009901 = 990 \times 10^5 + 09901 = 990^2 + 09901^2$$

The **type declaration** section consists of the word **type** followed by type declarations. A type declaration is the type identifier and its declaration, separated by an "=" symbol. Several standard type identifiers are predeclared in Pascal: integer, real, Boolean, and char. You don't have to declare these. You declare only types that you build from other types, or subranges of other types. Here are several examples:

```
type
  Number = Integer;
  10List = (Channel, Disk, Tape);
  Month  = (Jan, Feb, Mar, Apr, May, Jun, Jul, Aug, Sept,
            Oct, Nov, Dec);
  Day    = (Mon, Tue, Wed, Thu, Fri, Sat, Sun);
```

A **variable** represents a quantity that may take on different values during the execution of a program. When a variable is declared, computer memory is allocated, thereby providing a place to put a particular type of data. A variable declaration also provides a name for that place. Variable declaration consists of the word **var** followed by one or more identifiers, separated by commas, each followed by a colon and a type. This declaration creates a new variable of the specified type and associates it with the specific identifier. Here is an example of a variable declaration section:

```
var
  D, H, Number, R     :Integer;
  Root1, Root2        :Real;
  Flag                :Boolean;
```

The format for declaring variables is somewhat flexible. For example, the four integer type variables could also be declared like this:

```
D,
H,
Number,
R:Integer;
```

or this:

```
D           :Integer;
H           :Integer;
Number      :Integer;
R           :Integer;
```

Here are some variable declarations for the different data types

```
var
  Number1, Number2, Sum     :Integer
  Radius, Diameter, Area    :Real;
  One, Two, Three           :Character;
  Flag1, Flag2, Flag3       :Boolean;
```

Since a declaration does not have to appear on just one line, you could rewrite the variable declarations as

```
var
  Number1,
  Number2,
  Sum
  :Integer;
  Radius,
  Diameter,
  Area
  :Real;
  One,
  Two,
  Three
  :Character;
  Flag1,
  Flag2,
  Flag3
  :Boolean;
```

Procedure and Function Definitions After all of your variables have been declared, you must define any procedures or functions your program is to include. Procedures and functions are miniature programs that are used by the main program.

Statements The steps in the program (the instructions which actually express the algorithm) are written as statements in Pascal. Statements are executed by the computer in sequence, from the first onwards, except that some statements have the ability to modify this flow of control.

Main Program The main body of a Pascal program consists of the words **begin** and **end**, with any number of executable statements between these words. Execution always starts with the first statement after **begin** and proceeds to the last statement before **end.** There may be several **begin . . . end** pairs, however, only the final **end** is followed by a period. This is the only place where a period should follow **end.**

At this point we are ready to look at a complete Pascal program. The following example program causes the computer to find the sum of two numbers and print the result.

```
program Addition;
var
  A, B, C      :Integer;
begin
  write('Enter two numbers:');
  readln(A);
  readln(B);
  C := A + B;
  writeln('The sum is ', C)
end.
```

NUMBER THEORY TRIVIA

More than a hundred years ago, Catalan conjectured that the only two powers of integers which differ by exactly one are 3^2 and 2^3, i.e., 9 and 8. It is interesting to note that there are powers of integers that differ by 1, 2, 3, . . .

9 and 8 (i.e., $3^2 - 2^3 = 1$)
27 and 25 (i.e., $3^3 - 5^2 = 2$)
128 and 125 (i.e., $2^7 - 5^3 = 3$)
125 and 121 (i.e., $5^3 - 11^2 = 4$)

Where does this stop? Catalan's conjecture says that, for a difference of one, 9 and 8 is the *only* solution. Can you prove this conjecture? Can you write a program to produce powers of integers that differ by 5, 6, 7, 8, 9, and so on?

NUMBER THEORY TRIVIA

If any integer from 6693 to 7086, inclusive, is divided by 7874, the quotient will have the digit 7 in the 7th position. For example, 6793 divided by 7874 equals .8627127 and 7013 divided by 7874 equals .8906527.

Between the words **begin** and **end** there are five executable statements. Let's look at these statements, one at a time.

- The write statement causes the computer to write whatever is enclosed within the parentheses on the display screen. Upon execution of this statement, the message

  Enter two numbers:

  will appear on the screen.

- The readln(A) statement causes the computer to wait for input from the keyboard. Since A is an integer variable, it's waiting for you to enter an integer number. When you type in a number, then press the *Enter* key, the program assigns that number to the variable A in the parentheses.

- The readln(B) statement causes the computer to wait again. It's waiting for another number followed by pressing the *Enter* key.

- The next statement evaluates the expression A + B. In other words, it calculates the sum of the two numbers represented by variables A and B using the formula

  C = A + B

  Let's suppose that, when prompted by the program, you had keyed in the values 246 (for A) and 112 (for B). After execution of this statement, the variable C would contain the value 358.

- The writeln statement causes the message

  The sum is

  to show on the screen, followed by the value for C. For example, if you had indeed entered 246 and 112 for A and B respectively, then the message would be

  The sum is 358

For the moment, don't worry about how this program worked. This example, is included here only to illustrate the format of Pascal programs.

Semicolons And Compound Statements

Pascal programs are written in free-format. Statements may stretch over several lines and they may contain spaces as desired, except within identifiers, reserved words, and compound symbols such as ": = ". In order to distinguish one statement from the next one, a **semicolon** is used as a statement separator. The following program illustrates this:

```
program AreaOfRectangle;
var
  Length     :Real;
  Width      :Real;
  Area       :Real;
```

absolute
and
array
begin
case
const
div
do
downto
else
end
external
file
forward
for
function
goto
inline
if
in
label
mod
nil
not
overlay
of
or
packed
procedure
program
record
repeat
set
shl
shr
string
then
type
to
until
var
while
with
xor

Pascal reserved words

```
begin
  readln(Length);
  readln(Width);
  Area := Length * Width;
  writeln('The area is ', Area)
end.
```

The four statements of this program, between **begin** and **end** are separated by three semicolons. Notice that neither the **begin** nor the statement immediately before **end** are followed by semicolons; **begin** and **end** *are not* statements (they are reserved words), and the semicolon is only used in separating statements.

A group of statements bracketed by **begin** and **end** are said to form a **compound statement.** The statement part of a Pascal program is always a compound statement. Compound statements may be written in free-format. For example,

```
begin
  a := b;
  b := c;
  c := d
end
```

is a compound statement consisting of three statements. This compound statement could also be written as

```
begin a := b; b := c; c := d end.
```

Reserved Words

Pascal uses some special words for putting programs together. About forty-five of these words are termed **reserved words** or **keywords.** The meaning of these words cannot be changed. A second set is termed **standard identifiers.** These have also been given previous meanings. A list of both the reserved words and standard identifiers is found in the margins.

12.2 READING AND WRITING

Every sensible program needs to communicate in some way with the world outside the computer by reading in data or writing out the result. The power of a computer program comes from the fact that the same set of instructions can process different sets of data. In order for this process to occur, the data items are declared as variables. When the program is executed, the actual values are made available to the program. When the required results have been calculated, they are displayed on the computer screen or on a printer. This input and output involves transferring data from the device into the location assigned to the variable in the computer's memory or copying the contents of this location out to the device.

To simplify the input/output (I/O) process, Pascal includes four statements which control the input and output of data: read, readln, write, and writeln. The programmer does not have to be concerned

Constants
false
maxint
true

Types
boolean
char
integer
real
text

Variables
input
output

Functions
abs
arctan
chr
cos
eof
eoln
exp
ln
odd
ord
pred
round
sin
sqr
sqrt
succ
trunc

Procedures
dispose
get
new
pack
page
put
read
readln
reset
rewrite
unpack
write
writeln

Pascal predefined identifiers.

with how input and output is actually controlled; the person merely writes the name of the input or output statement and the name of the data item.

This section deals with the inputting of data from a keyboard and outputting of data to a visual display terminal (VDT) or printer. The VDT with associated input keyboard offers fast, convenient input and output for many kinds of computer interaction. Initial program entry and subsequent editing are particularly well-suited to the VDT and keyboard with their fast action and correcting facilities.

All of these input/output devices have one important characteristic in common: They are all character-based. In other words, all communication between keyboards, VDTs, printers, and the computer itself takes place in the form of sequences of characters in the character set of the computer. This kind of communication allows a consistent approach to be developed regardless of the physical nature of the input or output device attached to the computer.

All Pascal programs have a standard input channel, or "stream," called *input* and a standard output channel called *output*; both are organized entirely as lines of characters. The term "stream" stems from the idea of characters flowing one after the other from device to computer or vice versa. A program may read values from the input stream or write characters out onto the output stream using statements that require no insight into the physical nature of the device at the other end. It is a major function of the operating system of the computer to provide any necessary transformations between a generalized input/output scheme of this nature and the I/O devices, and to allow flexible reassignment of I/O devices. A simple change in an operating system command to send output from a Pascal program to a printer instead of a visual display terminal screen, without in any way needing to change the program itself, is an example of reassignment of I/O devices.

The Read Statement

In order for a computer to solve a problem, it must be provided not only with instructions telling it what to do, but also with data to use when carrying out the instructions. In Pascal, the read and readln statements may be used to supply data to a program. The effect of both statements is similar. For purposes of this presentation, the input medium is assumed to be a keyboard.

Let us consider the **read statement** first. Its general form is

read(Identifier 1, Identifier 2, . . ., Identifier n);

It consists of the word *read* followed by a list of identifiers which is enclosed in parentheses. The identifiers are separated by commas and the statement ends with a semicolon. As an example, consider

read(ALPHA, BETA, GAMMA, DELTA);

When the read statement is executed, values will be assigned to each identifier in the list, provided that the required data is supplied. In this example, three numbers must be supplied. The numbers must agree with the type of the identifier. Suppose that ALPHA and BETA had been declared as integers, while GAMMA and DELTA were declared as real. Then, a number with no decimal point must be entered for ALPHA and BETA, while numbers with fractional parts are entered for GAMMA and DELTA.

Now let us consider how the data would be entered. Four numbers are entered; the numbers can be separated by either spaces or (carriage) returns. Any number of spaces and new lines will be skipped over automatically in the input stream to find the start of a new number. Subsequent characters are taken as part of that number until a character is found which cannot be part of the number (e.g., a space). Thus, successive numbers must be separated by at least one space or new line. A return must always follow the last item of data. Several ways of entering the same data would be

```
88        162
40.2      3.98E2
```

or

```
88     162     40.2     3.98E2
```

or

```
88
162
40.2
3.98E2
```

Data are always read in order. For instance, in the previous read statement, the value assigned to ALPHA would be the first data encountered; data for BETA would be the second one; data for GAMMA would be next, etc.

More than one read statement can be in a program. For instance,

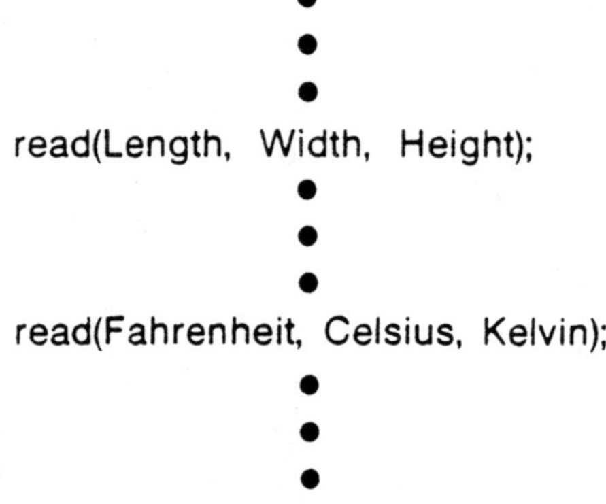

These two statements are equivalent to the single statement

```
read(Length, Width, Height, Fahrenheit, Celsius, Kelvin);
```

NUMBER THEORY TRIVIA

There are groups of three numbers, say L, M, and N, such that the divisors of L add up to M, the divisors of M add up to N, and the divisors of N add up to L. Groups of numbers with this chainlike property of divisor association, whether there are 3, 4, 5, or even more members in the group, are called **sociable groups**. An example of a five member group is

12,496
14,288
14,264
14,536
15,472

A computer search for sociable groups with 10 or fewer members has been made for all numbers up to 60 million. Ten groups were located in this search.

NUMBER THEORY TRIVIA

The sum of the first n numbers is $(n \times (n + 1)/2$. For example

$1 + 2 + 3 = 6$
$1 + 2 + 3 + 4 + 5 = 15$
$1 + 2 + 3 + 4 + 5 + 6 + 7 = 28$
$1 + 2 + \ldots + 1000 = (1000 \times 1001)/2$

Now six items of data can be entered on one to six lines. Any extra data (e.g., a Seventh item of data keyed in on the terminal) will be ignored. If there is not enough data, then an error results.

The Readln Statement

The **readln statement** operates in exactly the same manner as the read statement, except that it *reads* values for the specified variables and then goes on to the next line. The remaining characters in the line are skipped over. For instance, consider the following statements:

```
readln(A, B);
readln(X, Y);
```

The input data is

```
32    46    98    77    19
5.8   7.3   1.2   4.4   8.0
```

The values assigned to the variables are

Variable	Value
A	32
B	46
X	5.8
Y	7.3

On the first line of input, only the first two values, 32 and 46, are used; the values 98, 77, and 19 are skipped. On the second line of input, only the first two values, 5.8 and 7.3 are used; the values 1.2, 4.4, and 8.0 are skipped. The next read or readln statement in the program will commence reading on a new line.

To sum up, the read statement causes data to be read in the order in which it is entered with no regard to returns. The readln statement functions in the same way except that after a readln statement is executed, all the remaining data on the current line is ignored.

Write And Writeln Statements

To do even a simple problem, we usually need a way to get data into the computer and a method of printing the computed answer. Two ways of getting data into a computer were discussed in the last section. Results may be printed by using write and writeln statements. **Write and writeln statements** enable you to see the results of the executed program. For purposes of this presentation, the output medium is assumed to be a printed page, such as that generated by a display screen printer.

The basic form of the write statement is

```
write(output list);
```

where *output list* consists of one or more output values separated by commas. Each item in the list must give a value which is either

DID YOU KNOW?

Floppy disks exposed to extreme heat or cold will be ruined if used before they return to room temperature. *Solution:* Wait at least two hours after removing the disks from an extremely hot or cold place before inserting them in the computer's disk drive.

integer, real, character, Boolean, or a string of characters enclosed in single quotes. In many cases the items will be single variable identifiers, but any integer or real expression is permitted. If the output list is a literal, it is simply displayed. If the list item is a variable, the value of that variable is displayed. If the list item is an expression, the expression is evaluated and its value is displayed.

The writeln statement is essentially the same as the write statement except that, after all the data in its list are printed, the writeln statement causes the output device to go to a new line.

To illustrate the difference between the write and writeln statements, consider the following example. Assume that values for A, B, C, etc., have already been assigned earlier in the program (A := 10, B := 20, C := 30, D := 40, E := 50, F := 60, G := 70, H := 80, I := 90). The statements

```
        •
        •
        •
write(A, B, C);
write(D, E);
write(F, G, H, I):
        •
        •
        •
```

cause the computer to print

```
102030405060708090
```

Since the final statement is a write statement, the output device remains on the same line and the next write and writeln statements to be executed continue printing on the same line.

On the other hand, if you enter the following writeln statements instead of write statements

```
          •
          •
          •
writeln(A, B, C);
writeln(D, E);
writeln(F, G, H);
          •
          •
          •
```

the computer prints

```
102030
4050
60708090
```

and the next write or writeln statement begins printing on a new line.

We will use the writeln statement more frequently than the write statement, because we want each output statement to produce a single line of output.

Printing Alphanumeric Labels The write statement can print messages or strings which consist of letters, numbers, and/or special symbols. For example, the statement

writeln('Test Scores')

will cause the computer to print on the terminal whatever is between single quotation marks; therefore the above statement would cause the computer to print

Test Scores

A writeln statement containing single quotation marks will print the message *exactly*. For example, the statement

writeln('WINDSURFING IS FUN');

will cause the three words enclosed in single quotation marks to be printed with one space between each word. The statement

writeln ('WINDSURFING   IS   FUN');

will cause three spaces to appear between the words WINDSURFING and IS, and five spaces to appear between the words IS and FUN. In each case, the skipped spaces precisely match the skipped spaces in the string.

The following example shows how a computer prints information exactly as it is written in the program.

```
program SamplePrintProgram;
begin
  writeln('Good afternoon human');
  writeln('Punch my keys to find');
  writeln('out what I can do.')
end.
```

COMPILE/RUN

```
Good afternoon human
Punch my keys to find
out what I can do.
```

Note: This command will be used in several examples in this book. It means to compile and execute the program. Following this command is the output produced by the Pascal program.

Printing Values One form of the write statement is used to print values of variables. For instance,

•
•
•
write(ZEBRA, TIGER, ELEPHANT);
•
•
•

It is assumed that the values of ZEBRA, TIGER, and ELEPHANT have been assigned earlier in the program. When the writeln statement is executed, the numerical values of ZEBRA, TIGER, and ELEPHANT will be printed. Suppose that ZEBRA is an integer and that TIGER and ELEPHANT are real. In addition, their values are

ZEBRA = 822
TIGER = 46.3
ELEPHANT = 2.99E2

The output would appear as

```
822        4.6300000000E01        2.9900000000E+02
```

If there is enough room, the output of a write statement will be printed on a single line, as will the data resulting from a sequence of write statements. For instance, the following statements will cause output which is equivalent to that caused by the previous write statement:

```
write(ZEBRA);
      •
      •
      •
write(TIGER, ELEPHANT);
      •
      •
      •
```

The following three statements will also produce output which is identical to that produced by the two previous write statements.

```
write(ZEBRA);
write(TIGER);
write(ELEPHANT);
```

Mixing Alphanumeric Labels and Values Messages and variables can also be mixed in the same writeln statement. For example, the statement

```
writeln('TOTAL = ', T);
```

will cause the message TOTAL = and the value of T to be printed. Thus, if the current value of T is 625, then the following would be printed:

```
TOTAL = 625
```

No, diskettes are not cheerleaders for computer games.

Labeling Input Printing text can be very helpful when data are entered on a keyboard/terminal. For example, whenever data have to be entered into a program, the person running the program has to know what data are to be entered, and in what order and form. Appropriate printing of text can prompt the user.

Note how the following program uses write statements to supply information to the program user.

DID YOU KNOW?

You should copy important information on floppy disks to another disk every 18 months. Why? The oxide on floppy disks begins to deteriorate after two years and this could cause data to get lost. *Another possibility:* Store the data on a hard disk.

```
program InputExample;
var
  A, B       :Integer;
begin
  write('What is the value of A? ');
  readln(A);
  write('What is the value of B? ');
  readln(B);
  writeln('The sum of A + B is ', A + B);
end.
```

COMPILE/RUN

```
What is the value of A? 264
What is the value of B? 312
The sum of A + B is 576
```

The user knows exactly what data to enter and in what order because the first two write statements in the program are used to identify what data values are to be entered.

Performing Calculations With The Writeln Statement The writeln statement can be used to direct the computer to perform arithmetic calculations. For example, the writeln statement in the following program computes and prints the area of a rectangle. Suppose the sides of the rectangle were 8 and 4 feet.

```
program RectangularAreaExampleOne;
var
  Length, Width      :Real;
begin
  writeln('Enter length:');
  readln(Length);
  writeln('Enter width:');
  readln(Width);
  writeln(Length * Width);
end.
```

COMPILE/RUN

```
Enter length:
8
Enter width:
4
  3.2000000000E+0
```

The output is a result of the computation performed based on the third writeln statement in the program. Only the area of the rectangle is printed out.

```
HELLO
WHAT IS THE PASSWORD? ABC 12345
WHAT IS YOUR NAME? BOB SMITH
WHAT IS YOUR BILLING NUMBER? R123
SYSTEM READY
```

Input prompts and output descriptions make programs more "user friendly."

Print Headings Typically, the desired output has headings, making it more meaningful and easier to understand. The following program prints headings as well as performing the necessary area calculation.

```
program RectangleAreaExampleTwo;
var
  Length, Width      :Real;
begin
  writeln('Enter length:');
  readln(Length);
  writeln('Enter width:');
  readln(Width);
  writeln('LENGTH              WIDTH              AREA');
  writeln(Length, '  ', Width, '  ', Length * Width);
end.
```

```
COMPILE/RUN

Enter length:
8
Enter width:
4
LENGTH              WIDTH              AREA
8.0000000000E0  4.0000000000E0  3.2000000000E1
```

Skipping Lines If you wish to skip a line between the heading and the body of a printout, an extra writeln statement may be added as follows:

```
writeln('LENGTH              WIDTH              AREA');
writeln;
```

Field-Width Parameters Pascal provides a mechanism for controlling the spacing of printed items. In the statement

```
writeln(L : N);
```

the value of L is output right-justified in a field which is N characters wide.

Consider the statement

```
writeln(365 : 12);
```

This is the last time he will try to sell us a computer.

The value to be printed is 365; the field-width parameter is 12. The 12 specifies that the value 365 will be printed in a field 12 characters wide (i.e., 12 character positions on the printed line will be reserved for the value 365). The printed value will be preceded by nine blank spaces, so that the blank spaces together with the three digits of 365 will together occupy 12 character positions.

To get a real value printed in conventional notation, we must follow it with two field-width parameters. The first of these gives the width of the field in which the real number is to be printed. The second specifies the number of decimal places that will be printed. Thus, the statement

```
writeln(8.42 : 8 : 3)
```

prints the value 8.42 in a field eight characters wide and with three digits to the right of the decimal point:

NUMBER THEORY TRIVIA

Look at the equation

159 × 48 = 7632

which contains each of the digits 1, 2, 3, . . ., 9 once only. Can you write a program to generate other pairs of numbers whose product gives a sum which uses all the digits only once?

*bbb*8.420

where *b* is used to indicate a blank space.

The statement

writeln(2500.0 : 10 : 2, -4632.7 : 12 : 4);

causes the computer to print

*bbb*2500.00*bb*-4632.7000

or, as it would actually appear

2500.00   -4632.7000

We can use field-width parameters with strings as well as integers and real numbers. For example,

writeln('TIC' : 6, 'TAC' : 6, 'TOE' : 6);

produces

*bbb*TIC*bbb*TAC*bbb*TOE

or, as the line would actually look in the printout

TIC   TAC   TOE

12.3 PERFORMING CALCULATIONS

In the previous section, we learned how to assign data values to variable names with the read and readln statements. Another way in which we can assign values to variables is with the assignment statement, which is discussed in this section.

Arithmetic Operators

In Pascal, the symbols used to represent some mathematical operations are different from the symbols you learned in mathematics classes. For example, the symbol for multiplication we learned in elementary arithmetic was "x", "•", or was implied (ab), but in Pascal the symbol for multiplication is an *. Six times nine would be written as 6 * 9 in Pascal instead of 6 × 9. Consider (x + y) ÷ 2 as written in mathematical form. In Pascal (x + y) ÷ 2 is equivalent to (x + y)/2. The Pascal arithmetic operators are

Operation	Pascal Symbol
Addition	+
Subtraction	-
Multiplication	*
Division	/

Arithmetic Expressions

Several kinds of Pascal statements may contain expressions which are very much like algebraic expressions and cause the current values of the specified elements to be combined in the specified ways. An element in an expression may be a variable such as Sum or a constant such as 963.

The computer normally performs its operations according to a hierarchy of operations. The system will search through an arithmetic expression from left to right and do specific operations according to a specified pattern. The order of precedence of the arithmetic operators is

1. Multiplying operators: *, /, **div**, and **mod**
2. Addition operators: + and -

One can alter the order of operations by using parentheses in the formula. The parentheses have no effect on the formula itself other than to direct the order of operations. The computer will always find the innermost set of parentheses and evaluate the part of the formula it finds according to the order of precedence of operators. If there are no parentheses, the expression is processed from left to right.

Let us consider the expression

6 * X + 3 * Y

What value will this expression produce? The value will depend upon how you *evaluate* the expression. Here are four different ways you could evaluate it. Assume that X = 20 and Y = 8.

((6 * 20) + 3) * 8 = 984
(6 * 20) + (3 * 8) = 144
6 * (20 + 3) * 8 = 1104
6 * (20 + (3 * 8)) = 264

As you can see, the order of evaluation can make a big difference. Pascal will carry out all multiplication and division before any addition or subtraction. Therefore, according to Pascal, the correct evaluation is

(6 * 20) + (3 * 8) = 120 + 24 = 144

If both of the operands of the multiplying and adding operators are of the type integer, then the result is of type integer. If one (or both) of the operands is of type real, then the result is also of type real. In other words, in an expression such as

10 * 4.8

where one operand is integer and the other real, then the integer operand will be converted to real before the multiplication is performed. Thus we may consider the above as a reasonable abbreviation of

10.0 * 4.8

Note that despite the fact that the result of this expression is 48.0, a whole number, the result is still a real number.

Division is different because two integer operands produce a real

THE IMPORTANCE OF MATHEMATICS

Much of modern technology is based on mathematical research: computing, high energy physics, and even biotechnology. In fact, mathematics is so important that major defense programs require scientific breakthroughs that would probably be impossible without advances in mathematics. Such advances are considered so important by the Defense Department that it has set up the Mathematical Research Institute (a mathematical think tank) at Cornell University. The Institute brings mathematicians together to do basic research in applied analysis, physical mathematics, numerical analysis, computing, statistics and applied probability.

NUMBER THEORY TRIVIA

Karl Gauss, in his famous book *Disquisitiones Arithmeticae*, introduced a certain class of numbers — we will call them **Gauss numbers**. Gauss succeeded in finding nine of these numbers; they were

1, 2, 3, 7, 11, 19, 43, 67, 163

He conjectured that these nine were all that existed, but this he could not prove. Nor could anyone else, from 1800 until 1934. Then in 1934, H. Heilbronn and E.H. Linfoot proved that there are *at most ten* Gauss numbers! In 1966, Harold Stark, an American mathematician, showed that there are only nine Gauss numbers; the tenth one doesn't really exist.

result if the operator "/" is used. Thus, 10/4 gives 2.5 and 10/5 gives 2.0, a real number.

Two operation symbols must not be used in succession unless separated by parentheses. Thus the incorrect expression x * - y should be written x * (-y) or -y * x.

The expression,

(a + b) / 2

is equivalent to

$$\frac{a + b}{2}$$

Parentheses may be nested to any depth; for example,

(a - (b + c)) * d

Note that operators of the same precedence are evaluated from left to right, thus

a / b * 2

is equivalent to

$$\frac{2a}{b}$$

and not

$$\frac{a}{2b}$$

Here are a few examples of putting mathematical expressions into their Pascal equivalents.

Mathematical Representation	Pascal Representation
a - b	A - B
$-a \div b$	-A / B
ab - xy	A * B - X * Y
a + bc	A + B * C
$(a + b) \div c$	(A + B) / C
$x \times y + 7$	X * Y + 7
$b^2 - 4ac$	B * B - 4 * A * C
length × width	LENGTH * WIDTH
$2\pi r$	2 * Pi * r

The operators **div** and **mod** are designed specifically to operate on integers: A **div** B means divide A by B and truncate (lose the remainder).

For example, 22 **div** 4 is equal to 5 (remainder 2 lost), and 9 **div** 5 gives the result 1. A **mod** B means A modulo B (the remainder after **div**).

NUMBER THEORY TRIVIA

In 1848, de Polignac conjectured that

> Every odd number is the sum of a prime and a power of 2.

For example:

$$25 = 17 + 2^3$$

Choose some odd numbers at random and verify this statement. Is the representation unique?

Remember when you first learned division? When you divided 5 into 13, you didn't get 2.6. You got 2 (the quotient) with a remainder of 3. Why? You were doing integer division, which required that the results also be integers. Well, that's what **div** and **mod** do: a **div** b returns the quotient, while a **mod** b returns the remainder. Here are more examples:

10 **div** 5 = 2	10 **mod** 5 = 0
11 **div** 5 = 2	11 **mod** 5 = 1
12 **div** 5 = 2	12 **mod** 5 = 2
13 **div** 5 = 2	13 **mod** 5 = 3
14 **div** 5 = 2	14 **mod** 5 = 4
15 **div** 5 = 3	15 **mod** 5 = 0

Two other points to keep in mind about **div** and **mod.** First, if a is less than b, then a **div** b will equal 0. In other words, while 300 **div** 300 equals 1, 299 **div** 300 equals 0. Second, if b equals 0, then both a **div** b and a **mod** b will result in some sort of error when they are executed.

Look at the following example. How would you evaluate the expression a **div** 3 * b? Assume that a = 30 and b = 12:

(30 **div** 3) * 12 = 120
30 **div** (3 * 12) = 0

If you chose the top one, you would be right. When operators of equal precedence show up, you take them in the order they come in: left to right.

The Assignment Statement

The most fundamental of all statements is the **assignment statement.** It is used to specify that a certain value is to be assigned to a certain variable. Its form is

variable := expression

The ":=" is a compound symbol signifying assignment of the value on the right-hand side to the variable on the left-hand side. In other words the expression on the right-hand side is evaluated, and the resulting value is copied into the location whose identifier is on the left-hand side.

The variable is a variable name; the expression may be a constant, a variable, or an arithmetic expression. The value of the expression is first computed according to the rules of arithmetic operations, and the result is assigned to the variable that appears to the left of the := symbol. Only one variable can be assigned a value at a time, and the := symbol is not the same as an equal sign. You may think of the := symbol as a "replaced by" symbol. Therefore, the statement is interpreted to mean "the value of the expression on the right of the *replaced by* sign replaces the current value of the variable on the left."

NUMBER THEORY TRIVIA

635,318,657 is the smallest integer which can be expressed as the sum of two fourth powers in two different ways is

$$635{,}318{,}657 = 133^4 + 134^4 = 59^4 + 158^4$$

Can you write a program to produce additional numbers of this type?

Assignment is possible to variables of any type as long as the type of variable and expression are the same. As an exception, if the variable is real, the expression may be integer.

A common use for the assignment statement is to assign some initial data to a variable, which is quite often used by other statements within the program. Here are a few examples.

```
x := 1
Length := 120
Number := 260
Angle := 30
```

In the first example, the statement causes the value 1 to be assigned to the variable named x. This takes the place of any previous value which x may have had, and this value will never change unless x is used to the left of the := symbol in another assignment statement or in read or readln statements. Thus, x may be used for comparison purposes or as part of an arithmetic expression in another assignment statement, and it will not alter its value. Similarly, in the next statement, the variable Length is assigned the value 120. In the next statement, the variable Number is assigned the value of 260, and in the last statement, the variable Angle is assigned the value of 30.

Here are some examples of assignment statements involving arithmetic expressions.

```
x := y
r := (a - b) / c
y := a + b * c
```

In the first example, the value in the storage location with the variable name y is stored in the location with the variable name x. The new value being stored in x replaces any value previously stored in that storage location. In the second example, the value of b is subtracted from the value of a. The difference is then divided by the value of c. The result is stored in the storage location with the variable name r. Any previous value for r is replaced by the result of this calculation. In the third example, the value in the storage location with the variable name c is multiplied by the value of b; this result is added to the value of a. The result is stored in the storage location with the variable name y.

As shown in the following examples, an assignment statement can also be used to clear storage locations for computations.

```
Amiga := 0
Atari := 0
Apple := 0
```

The constant of zero (0) is stored in storage locations represented by Amiga, Atari, and Apple. Any data previously stored in the same storage locations will be replaced by zeros. The storage locations are cleared of any old data.

NUMBER THEORY TRIVIA

Shown below is a pair of numbers whose squares add up to a cube number:

$$47^2 + 52^2 = 17^3$$

Use a microcomputer to investigate other sets of numbers.

$$a^2 + b^2 = c^3$$

There are eight such sets where *a*, *b*, and *c* are all less than 50.

The assignment statement can also be used to assign string data to a variable. For example, the statement

Name := 'Roger Harwood'

causes the string constant Roger Harwood to be assigned to the variable called Name.

The expression

Address := '491 Third Street'

causes this address to be assigned to the variable named Address.

It is quite common to write an assignment of the form

Count := Count + 1

Here, the action of the assignment is to evaluate the expression on the right-hand side (1 is added to the current value of the variable named Count) and to place this value in the location identified by the left-hand side (which happens to be the variable Count again). The effect is to increment the value of Count by 1.

Consider a program to read in a value of x and compute the value of the polynomial

$$y = x^3 + 4x^2 + 3x + 7$$

A statement to do the required computation is

```
y := x * x * x + 4 * x * x + 3 * x + 7
```

The program for this example follows.

```
program Polynomial;
var
  x, y      :Real;
begin
  write('Enter a value for x: ');
  readln(x);
  y := x * x * x + 4 * x * x + 3 * x + 7;
  writeln('x is ', x:8:2 ,'      y is ', y:8:2)
end.
```

```
COMPILE/RUN

Enter a value for x:8
x is     8.00      y is      799.00
```

Boolean Operations

Just as you can build up arithmetic expressions by combining the operators +, –, *, /, **div**, and **mod** with real and integer constants and variables, you can combine Boolean operators with Boolean constants and variables to build up Boolean expressions. There are three Boolean operators: **and**, **or**, and **not**. The operator's **and** and **or** are binary operators (i.e., they are used in conjunction with two operands). **Not**, on the other hand, is a unary operator, taking a

NUMBER THEORY TRIVIA

Seven-digit **plateau palindromic primes** are prime numbers of the form *abbbbba*. For example,

1333331
1444441
1777771
3222223
3444443

Can you write a program to compute other seven-digit plateau palindromic primes?

single operand. Since there are only two variables, True and False, that operands and results can take on, the effects of Boolean operators are commonly defined by the following table.

Possible Combination Of Values And Their Result

x	True	True	False	False
y	True	False	True	False
x **and** y	True	False	False	False
x **or** y	True	True	True	False
not x	False	False	True	True

Just as we can declare variables to be real or integer, we can declare variables to be Boolean:

```
var
  Red, Black, Green: Boolean;
```

You can also assign the values of True or False to Boolean variables by means of assignment statements. For example

```
Red := True;
Black := False;
Green := Black;
```

Here, the value of the Boolean variable Red is set to True, and the value of the Boolean variable Black is set to False; therefore, the value of the Boolean variable Green is set to False by assigning it the value of Black.

If you want to print the value of Red, you can write

```
writeln(Red)
```

and what will be written is

```
True
```

Boolean variables are used mostly to represent conditions. For example, the variable MOTOR might represent the informally stated condition "motor is on," in the sense that if the value of MOTOR is True, the motor is on, while if the value of MOTOR is False, the motor is off.

As shown above, Boolean variables can represent conditions. Thus X might mean the MOTOR IS ON and Y might mean DOOR IS CLOSED. Then X or Y represents the condition MOTOR IS ON or DOOR IS CLOSED. This condition is True whenever *either* the motor is on *or* the door is closed (i.e., whenever *either* (or both) of the subconditions is true). Looking at the previous table, you will see that x **or** y is False, if and only if *both* x and y are False. Similarly, for x **and** y to be True, both x and y must be True. Finally, **not** x (in this example) would mean "the motor is not on," which is False when the motor is on and True when the motor is off.

=	Equal to
<>	Not equal to
<	Less than
>	Greater than
<=	Less than or equal to
>=	Greater than or equal to

Relational operators.

Given an integer variable TEMPERATURE we may write an expression such as

TEMPERATURE > 0

which will be True or False according to the value of the variable TEMPERATURE. The relational operators, shown in the margin, work on all standard data types: integer, real, char, and Boolean. Operands of the type integer and real may be mixed. The type of the result is always Boolean (i.e., True or False). For example

a = b	True if a is equal to b
a <> b	True if a is not equal to b
a > b	True if a is greater than b
a >= b	True if a is greater than or equal to b
a < b	True if a is less than b
a <= b	True if a is less than or equal to b

By combining these relational operators with the logical operators (**and**, **or**, and **not**) we may develop quite complex Boolean expressions such as

ICE := (RAIN **or** SNOW) **and** (TEMPERATURE < 32)
NICEDAY := SUNNY **and** (**not** RAIN) **and**(TEMPERATURE > 70)

being careful to watch the precedence of each of these operators and to include parentheses to override undesirable ones.

The precedence of Boolean/relational operators is as follows:

Operator	**Precedence**
not	First
and	Second
or	Third
> >= < <= <> =	Fourth

Look at the following Boolean expression.

(A **and** B) **or** (C **and** (**not** A))

Let us see how this evaluates when A is False, B is True, and C is True. First, (A **and** B) is False, because A is False. Next, (**not** A) is True, because A is False. Because both C **and** (**not** A) are True. (C **and** (**not** A)) is True. Finally, the complete expression, being an **or** of two operands of which one is true, is True.

Extra parentheses were used in the previous example to clarify the expression. It can be equally well written as

A **and** B **or** C **and** **not** A

If this looks confusing to you, use the parentheses.

Comments

Suppose you wrote a program and put it away for use at some future date. Several months later, when you picked up the program again, you may not remember what the program does, or why you wrote it in the first place.

The purpose of using a high-level language like Pascal is to allow you to write down program instructions which are meaningful, thereby making it easier to write the program and also to read through it to understand how it processes the data. However it would also be convenient to be able to add some general information to the program to help jog your memory, or perhaps tell you what the program is all about. Pascal allows you to insert **comments** into your program.

Comments may be placed anywhere in the program. They offer you a convenient means to call attention to important variable names, to provide operating instructions, and to distinguish the major logical segments of a program. These comments are only for the use of the programmer and are not used at all by the machine.

Comments in Pascal are set off by curly braces

```
{ this is a comment }
```

or by using parentheses and asterisks

```
(*This is also a comment*)
```

A comment can be as long or as short as needed. It can contain any combination of characters, including reserved words, except for the characters that indicate the end of the comment. When the compiler finds the character(s) "{ or (*" it stops translating the program and scans over the subsequent characters until the character(s) "} or *)" is found. It then continues translating the program.

Comments should be used to describe the purpose of the program, state operating instructions, and explain any steps in the program that are not obvious. Comments should be brief and should only be used to add to the documentation of the program.

If a comment requires more than one line, a subsequent comment statement may be used, such as

```
(****Perfect Numbers****)
(****This program computes and****)
(****prints the first four perfect numbers****)
```

Here you can see that the entire comment requires three lines. Many programmers use comments to start each program. For example, the statements

```
(****Palindromic Numbers****)
(****Programmed by William Oates****)
(****December 14, 1990****)
```

NUMBER THEORY TRIVIA

An intriguing product is

16583742 × 9 = 149253678

where all the digits occur once on each side of the equality sign. Can you write a program to find any other products with this property?

identify the program name (Palindromic Numbers), the programmer's name (Willian Oates), and the date the program was written (December 14, 1990).

Comments are often used to identify variables used in a program. For example, these statements might be used to identify the variables in a program to compute the average height of several values.

```
(****AverageHeight****)
(****H — Height****)
(****S — Sum of heights****)
(****A — Average height****)
(****I — Looping counter****)
```

12.4 CONDITIONAL STATEMENTS

The If Statement

You may want to execute a statement(s) if a given condition prevails, and execute a different statement if that condition does not arise. The **if statement** is a conditional statement that enables you to accomplish this.

The general form of the **if** statement is

if Boolean expression **then** statement **else** statement

When an **if** statement is executed, the Boolean expression is evaluated first. If the value is True, the statement following **then** will be executed. If the value of the Boolean expression is False, the statement following **else** will be executed. In either case, the next statement executed will be the one following the **if** statement.

A simple form of the **if** statement uses only the **then** option. Its form is

if expression **then** statement

The Boolean expression itself may be as complicated as desired, using the logical operators **and**, **or**, and **not** where necessary to combine elements.

NUMBER THEORY TRIVIA

The last two **prime years** were 1979 and 1987. The next one will be in 1993.

NUMBER THEORY TRIVIA

The Russian mathematician L. Schnirelmann (1905-1938) proved that every positive integer can be represented as the sum of not more than 300,000 primes (A step in the right direction?). Later another Russian mathematician, I.M. Vinogradov, proved that there exists a number N such that all numbers larger than N can be written as the sum of, at most, four primes.

Several sample **if** statements follow.

```
if speed = 100 then writeln('Speed equals 100 mph');
if C < A then counter := counter + 1;
if a = b and c > d then s := 0;
if sky > 42 then writeln('Star 231 found');
```

The following statement is also a valid **if** statement.

```
if
  hours > 40
then
  grosspay := 40 * rate + (hours - 40) * rate * 1.5
else
  grosspay := hours * rate;
```

Either the **then** option or the **else** option is executed but not both.

Because the activity associated with a comparison's outcome is expressed as a Pascal statement, the activity can be simple or extensive. When an activity cannot be expressed as a simple statement, you may specify it as a sequence (a compound statement), and the syntax rule for the **if** statement still is obeyed. To explore this decision structure further, consider the following example in which a variable named speed is compared to another variable named maximum.

```
if
  speed = maximum
then
  begin
    statement;
    statement;
    . . . . .
    . . . . .
    statement;
    statement;
  end
else
  statement;
```

If the two values are equal (i.e., if the outcome of the comparison is true), the program will execute the entire sequence specified within the compound statement. If not, that statement will be ignored and the program will execute the alternative simple statement instead. Indentation of the compound statement emphasizes the fact that, in concept, the compound statement is a single activity related to one of the two possible outcomes of the true-or-false test.

Consider the following program. A linear equation in one unknown can be solved by this program. The program uses an **if** statement to branch to the **else** warning whenever A = 0 (this eliminates a machine division by 0 in the case in which coefficient A of AX + B = C is 0).

```
program LinearEquation;
(****Program solves the equation AX + B = C****)
var
  A, B, C       :Integer;
  X             :Real;
begin
  write('Enter the coefficients A, B and C : ');
  readln(A, B, C);
  if A <> 0
    then
      begin
        X := (C - B) / A;
        writeln('X =    ', X:6:2)
      end
    else
      writeln('Not a Linear Equation because A = 0');
end.
```

The Case Statement

Choosing between one of several alternative control paths is critical to programming computers. You've seen how the **if** statement in its simplest form can choose between two alternatives based on the value of a Boolean expression. By nesting **if** statements one within another, you can choose among several different control paths. The problem of readability appears, however, when you nest more than two or three **if** statements. The **if** statement was designed for making decisions between two alternatives. When more alternatives are introduced, the **if** statement becomes clumsy.

The **case statement** is a generalization of the **if** statement, and it enables the process to execute one of several actions according to the value of a selector. It consists of an integer variable (the selector) and a list of statements, each preceded by a case label of the same type as the selector. It specifies that the statement whose case label is equal to the current value of the selector to be executed. If none of the case labels contain the value of the selector, then no statement is executed.

All my dreams are in Pascal.

A case label consists of any number of constants or subranges separated by commas and followed by a colon. A subrange is written as two constants separated by the subrange delimiter ". .". The type of the constants must be the same as the type of the selector. The statement following the case label is executed if the value of the selector equals one of the constants, or if it lies within one of the subranges.

The general form of the **case** statement where a value is a list of values separated by commas would be written in this manner:

```
case integer variable of
  value 1 : statement 1;
  value 2 : statement 2;
  value 3 : statement 3;
  value 4 : statement 4;
            •
            •
            •
  value n : statement n;
end; (****of case statement****)
```

As an example consider a program to calculate the number of days in a given month. (Assume the year is not a leap year!) Note that just one of the writeln statements will be executed, depending on the value of the variable Month.

```
program DaysInMonth;
var
  Month, Days       :Integer;
begin
  write('Enter number of month: ');
  readln(Month);
  case Month of
    2:writeln('Days = 28');
    4,6,9,11:writeln(Days = 30');
    1,3,5,7,8,10,12:writeln('Days = 31');
  end; (****of case statement****);
end. (****of program****)
```

COMPILE/RUN

```
Enter number of month: 8
Days = 31
```

NUMBER THEORY TRIVIA

A **sum number** is one which conforms with

$$N = 10^n x + y$$
$$= \Sigma x + \Sigma y$$

were $\Sigma x = x(x + 1)/2$ is the sum of the integers 1 to x, and $\Sigma y = y(y + 1)/2$ is the sum of the integers 1 to y. For example $\Sigma 6 = 6(7)/2 = 21 = 1 + 2 + 3 + 4 + 5 + 6$.

For n = 3 there are 5 sum numbers. One of them is

$$90\,415 = \Sigma 90 + \Sigma 415$$

For n = 4 there are 29 sum numbers. One of them is

$$120\,1545 = \Sigma 120 + \Sigma 1545$$

For n = 5 there are 11 sum numbers. One of them is

$$420\,09156 = \Sigma 420 + \Sigma 9156$$

Can you write a program to find all the sum numbers for n = 3? For n = 4? For n = 5?

In the following program a number between 1 and 7 is read in and converted on output to the word for the appropriate day of the week (i.e., Monday, Tuesday, Wednesday, etc.). When Dayno is 1, only the statement with the value 1 is selected; when Dayno is 2, the statement with the value 2 is selected; and so on.

```
program DayOfWeek;
var
  Dayno      :Integer;
begin
  write('Enter number of day:  ');
  readln(Dayno);
  case Dayno of
    1:writeln('Monday');
    2:writeln('Tuesday');
    3:writeln('Wednesday');
    4:writeln('Thursday');
    5:writeln('Friday');
    6:writeln('Saturday');
    7:writeln('Sunday');
  end; (****of case statement****)
end. (****of program****)
```

```
COMPILE/RUN

Enter number of day: 5
Friday
```

The previous program could be represented using **if** statements; however, the **case** statement is more concise and clear. When the decisions in a group of nested **if** statements are mutually exclusive (i.e., if any one being true implies that the rest are false), then a **case** statement is probably the appropriate representation.

Explicit Transfer Of Control

Pascal includes an unconditional branch statement for programming situations in which you always want to divert program flow from the next sequential statement. This statement does not test a condition or counter, but always proceeds to the program line prefaced with the appropriate label.

Unlike the other control statements, the **goto statement** is not at all vital to Pascal programming. With good planning and design and Pascal's range of control statements, you should be able to program for every contingency without using any unconditional branches.

NUMBER THEORY TRIVIA

Who did not contribute to the theory of numbers?

a. Pascal
b. Euler
c. Fibonacci
d. Fermat
e. Eratosthenes
f. Garfield
g. Slowinski
h. Leibniz

Nonetheless, the goal of programming is to solve problems as easily as possible. If you find it convenient to use a **goto** statement, use it. Eventually, however, you will begin to think in terms of Pascal's control structures and find yourself never using the **goto** statement.

A **goto** statement consists of the word "goto" followed by a label identifier. It serves to transfer further processing to that point in the

program text which is marked by the label. The general form of the **goto** statement is

goto label;

where "label" is an integer number or an identifier.

12.5 LOOP CONTROL

Conditional execution adds a lot to your programming capabilities, but you still can only execute each statement once. Because you need the ability to repeat a section of code a number of times or until some condition is met, a process known as *repetition* or *iteration* can be implemented.

Pascal uses three statements to repeat program processes: for, while, and repeat.

A **loop** is a program segment designed to be executed repeatedly during each execution of the program. The creation of a loop involves three steps beyond those required by nonrepetitive programs: (a)*initialization* (putting various variables in the proper state for starting execution of a loop); (b) *statement modification* (to select the statements to be executed during each repeat operation); and (c) *index modification and testing* (to record the number of times the loop has been traversed since its most recent initialization, and to exit from it when it has been traversed the required number of times).

Loops can be nested (i.e., contain other loops) to a depth usually limited only by the computer's storage capacity. The number of times an inner loop is repeated is the product of its own repeat count and those of all the loops it nests within, therefore, the code within a deeply nested loop has enormous leverage for good or bad because it may be executed thousands or millions of times in a single run of the program. Since it is common for a sizable program to be one big loop containing many multilevel nests of loops, the question of creating and optimizing loops is part of good program design.

Let us now look at the Pascal statements that are used to perform looping.

The For Loop

There are situations where you want to have a statement or group of statements executed a number of times, or you want to know how many times you have executed a group of statements. You may even want a variable to step through a range of values, executing a statement (or group of statements) once for each value. The **for statement** allows you to do this.

The general form of the **for** statement is

for Counter := Expression 1 **to** Expression 2 **do** statement;

NUMBER THEORY TRIVIA

Integers such as 1729 and 40,033 can be expressed as the sum of two cubes in two different ways:

$$1729 = 9^3 + 10^3 = 1^3 + 12^3$$

$$40{,}033 = 16^3 + 33^3 = 9^3 + 34^3$$

Can you write a program to produce other numbers of this type?

The form of a **for** statement with compound statement is

```
for Counter := Expression 1 to Expression 2 do
  begin
    statement 1;
    statement 2;
    statement 3;
         •
         •
         •
    statement n;
end;
```

Expression 1 is the lower bound and Expression 2 is the upper bound of the range for counting. The variable Counter, which serves as a counter for the loop and is a variable that must be declared, is assigned to Expression 1 when the loop is encountered and is then automatically incremented each time the loop is executed until Counter equals Expression 2; then the execution of the **for**-loop halts. If Expression 1 is greater than Expression 2, the statement is never executed. The variable Counter may be of any type except type Real; the loop limits must be of a compatible type.

The **for** statement can be used to simplify the formation of loops. Consider this program.

```
program Looping;
var
  Index      :Integer;
begin
  for Index := 1 to 8 do
    writeln('My name is Tom Wilson');
end.
```

In this program, Index is the name of a counter; it has been set to vary from 1 to 8. The loop consists of one statement, and the program prints "My name is Tom Wilson" eight times.

The following program prints the value of the counter as the loop is executed ten times.

```
program Loop;
var
  LoopCounter      :Integer;
begin
  for LoopCounter := 1 to 10 do
    writeln('Counter = ', LoopCounter);
end.
```

Look at the following program. The number of passes through the loop is known before the execution of the loop has been started but not before the execution of the program has begun. The product of all integers from 1 through N is called N-factorial, and denoted by N! Thus,

$$N! = N(N - 1)\ (N - 2)\ \ldots\ 3 \times 2 \times 1$$

In particular, we have

$$1! = 1$$
$$2! = 2 \times 1 = 2$$
$$3! = 3 \times 2 \times 1 = 6$$
$$4! = 4 \times 3 \times 2 \times 1 = 24$$

Proceeding in this way, we can make a table of N!. The task soon becomes laborious, because the factorials increase in size at a fantastic rate. The number of permutations of the letters of the alphabet 26!, is greater than 4×10^{26}.

```
program Factorial;
var
  N, Nfactorial, I        :Integer;
begin
  write('Enter N:   ');
  readln(N);
  Nfactorial := 1;
  for I := 2 to N do
    Nfactorial := Nfactorial * I;
  writeln(N, ' -Factorial is ', NFactorial);
end.
```

Suppose you wanted to write a program to determine the square of each integer from 1 to 100. Without using a loop, the program would require over 100 separate statements, and most of these statements would repeat the process to determine the square of an integer value.

To avoid the difficult task of writing the same statement a number of times, we can write a simple program that will provide the same function; that is, build a loop which will count numbers 1 through 100, and print the square of each number. Here is an example of how to do this.

```
program SquareIntegers;
var
  Number      :Integer;
begin
  writeln('NUMBER', '   ', '   SQUARE');
  writeln('----------------------------');
  for Number := 1 to 10 do
    writeln(Number, '                          ', Number * Number);
end.
```

Consider the following problem. Four scales measure temperature: Fahrenheit (F), Celsius (C), Kelvin (K), and Rankine (R). Fahrenheit temperatures are converted to Celsius by subtracting 32° and multiplying the difference by 5/9. Kelvin temperatures are obtained by adding 273° to the Celsius reading. Rankine temperatures are obtained by adding 460° to the Fahrenheit reading. The following program computes temperature equivalents of the following

Fahrenheit temperatures: 100°, 110°, 120°, 130°, 140°, 150°, 160°, 170°, 180°, 190°, 200°, 210°, 220°, 230°, and 240°.

```
program TemperatureConversion;
var
  Fahrenheit, Rankine, Counter      :Integer;
  Celsius, Kelvin                   :Real;
begin
  Fahrenheit := 90;
  writeln('FAHRENHEIT    CELSIUS    KELVIN    RANKINE');
  writeln('--------------------------------------------------');
  for Counter := 1 to 15 do
    begin
      Fahrenheit := Fahrenheit + 10;
      Celsius := 5.0/9.0 * (Fahrenheit - 32);
      Kelvin := Celsius + 273.0;
      Rankine := Fahrenheit + 460;
      writeln('    ', Fahrenheit, '    ', Celsius:8:2, '    ',
                  Kelvin:8:2, '    ', Rankine);
    end;
end.
```

After printing the heading, the program computes and prints Celsius, Kelvin, and Rankine values for 15 different values of Fahrenheit.

Nested Loops Nested loops occur when it is necessary to repeat a loop more than one time. Whenever we have one **for**-loop contained entirely within another **for**-loop, these loops are called **nested loops**. here is an example of nested loops.

```
program NestedLoops;
var
  A, B      :Integer;
begin
  writeln('A', '        ', 'B');
  writeln('—          —');
  for A := 1 to 4 do
    for B := 1 to 3 do
      writeln(A, '        ', B);
end.
```

This program contains two loops. The outer loop is controlled by the first **for** statement. The inner loop is controlled by the second **for** statement. The inner loop cycles three times for each cycle of the outer loop.

Let us now look at a program that prints the following pattern:

```
*
* *
* * *
* * * *
* * * * *
* * * * * *
```

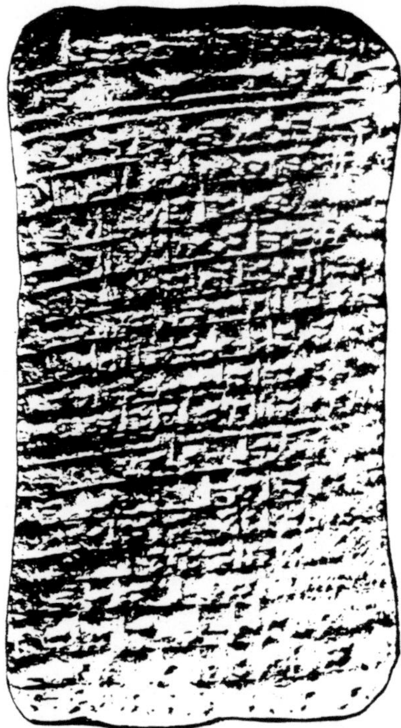

A Babylonian clay tablet of cuneiform script, dating from 1800 B.C., offers a procedure for figuring compound interest. The algorithm is expressed as a specific example, determining how many years and months it takes to double a certain quantity of grain called a *kur* at an annual interest rate of 20 percent (the grain served as currency). A Pascal program can present the same procedure in terms a computer can understand.

```
program AncientCalculation;
var
  Years                          :Integer;
  Rate,Grain,Previous            :Real;
  MonthGain,MonthNumber          :Real;
begin
  (*Set interest rate to 20 percent*)
  Rate : = 0.20;
  (*After one year the total is the*)
  (*original kur of grain plus*)
  (*one year's interest payment*)
  Years := 1;
  Grain := 1.0 + Rate;
  (*Continue calculating the amount*)
  (*of grain after each year until*)
  (*the goal of two kurs is*)
  (*met or exceeded*)
  while Grain < 2.0 do
    begin
      Previous := Grain;
      Grain := Grain * (1.0 + Rate);
      Years := Years + 1;
    end;
  (*The goal of two kurs of grain*)
  (*was reached during the year*)
  (*before the current value of Years.*)
  (*Next, calculate the amount*)
  (*of interest month-by-month*)
  (*within the last year.*)
  MonthGain := (Grain - Previous)/12.0;
  MonthNumber:=(2.0-Previous)/MonthGain;
  Writeln('The grain will be doubled');
  Writeln('to two kurs after ', Years - 1);
  Writeln('years and', MonthNumber);
  Writeln('months.');
end.
```

Two **for** statements are needed to print this pattern — one to print the six lines and one to print the proper number of asterisks on each line. The second **for** statement is part of the first. The following program uses nested **for** statements.

```
program TriangularPattern;
var
  I, J       :Integer;
begin
  for I := 1 to 6 do
    begin
      for J := 1 to I do
        write('*');
      writeln;
    end;
end.
```

For Statement with Downto The **for** statement always uses an increased value of the control value for the next pass through the loop. It is sometimes useful to begin with a high value and count down to a lower value. If you want to decrease the variable Counter instead of incrementing it, you can use the format

for Counter := Expression 1 **downto** Expression 2 **do**

Now, the variable Counter decreases each time until it is less than Expression 2, and the statement is never executed if Expression 1 is less than Expression 2.

The While Loop

Because the **for** statement cannot easily meet all of one's repetitive needs, Pascal has two other statements, both controlled by Boolean expressions. The first is the **while statement.**

The general form of the **while** statement is

while Boolean expression **do** statement;

The form of a **while** statement with compound statement is

```
while Boolean expression do
  begin
    statement 1;
    statement 2;
    statement 3;
        •
        •
        •
    statement n;
  end;
```

The Boolean expression can be an expression that evaluates to a Boolean value, such as K < 100, or it can simply be a Boolean variable. The statement is repeatedly executed as long as the expression is True. If the value is False, the statement is not executed at all.

Assuming that the Boolean expression came out True the first time, the code goes back and evaluates the Boolean expression after executing the statement. If it is still True, the statement is executed again. If it is False, the **while**-loop ends and control passes on to the next statement in the program. In short, as long as the Boolean expression is True, the statement will be executed repeatedly. The loop will end only when the expression comes up False.

The most important property of a **while** statement is that its Boolean expression is tested *before* the statement is executed. A corollary to this is that there are cases when the statement will never be executed at all. Keep this in mind as we discuss the **repeat** statement in the following section.

The **while**-loop is illustrated by the following program. Integer variable Number is initialized with a value of 1 in line 8, and incremented by line 12 each time the **while**-loop of lines 9 through 13 is executed. While Number is less than or equal to the maximum value MaxNumber input by the user in the execution of line 6, the two active lines of the **while**-loop (lines 11 and 12) continue to be executed. The last pass through the loop occurs when MaxNumber = 13, and the program ends at line 14.

```
program Count1;
var
  Number, MaxNumber       :Integer;
begin
  write('Count up to what number — inclusive?');
  readln(MaxNumber);
  writeln;
  Number := 1;
  while Number <= MaxNumber do
    begin
      writeln(Number:6);
      Number := Number + 1;
    end; (****of while-loop****)
end. (****of program Count1****)
```

NUMBER THEORY TRIVIA

Five-digit **plateau palindromic primes** are prime numbers of the form *abbba* (Note: numbers of this form are divisible by 3 when *a* = 3, 6, or 9). For example,

13331
15551
16661
19991
72227

Can you write a program to produce other five-digit plateau palindromic primes?

The Repeat Statement

The **repeat** statement can be used when you need a loop that will always execute at least once. The condition test for this statement is not performed until the statement part has been executed. The general form of the **repeat** statement is

repeat statement **until** Boolean expression;

The form of a **repeat** statement with compound statement is

```
repeat
  statement 1;
  statement 2;
  statement 3;
  statement 4
        •
        •
        •
until Boolean expression;
```

It may appear that there are several statements between **repeat** and **until**, however, they are considered to make up a single compound statement even though the traditional **begin** and **end** markers are not required.

Repeat-loops are very similar to **while**-loops in three important ways. First, the statements within the loop are *always* executed at least once (i.e., all the statements in the loop are executed before the Boolean expression is resolved). Second, the statements continue to execute as long as the Boolean expression is False. Third, the **repeat**-loop can directly execute more than one statement and, therefore, doesn't need the **begin** and **end** markers to handle multiple statements.

One common use of **repeat**-loops is to condition input, (i.e., to prompt the user for some value and to continue to prompt until the value entered is one of those that you decide to allow).

A **repeat**-loop is illustrated in the following program, which prints out integers up to any desired value.

```
program Count2;
var
  MaxNumber, Number      :Integer;
begin
  write('Count up to what integer? ');
  readln(MaxNumber);
  writeln;
  Number := 1;
  repeat
    writeln(Number);
    Number := Number + 1;
  until Number > MaxNumber;
end.
```

The variable Number is initialized to a value of 1 in line 8. Lines 10 and 11 are repeated until Number has assumed a value greater than the variable MaxNumber, the desired value input by the user by line 6 when the program is executed. Line 11 adds 1 to the previous value of Number and makes this larger sum the new value of Number.

Apparently, Professor Erk, our notions about the state of ancient science will have to undergo some revision.

Choosing Loop Structures

Choosing the correct loop structure to use in any application depends largely on the conditions under which that loop is to be repeated. The initial choice is among a **for**, **while**, or **repeat** statement, since these are used in very different circumstances. If the number of repetitions is known beforehand, the **for** statement is usually the appropriate statement to express the situation; otherwise, the **while** or **repeat** statement should be used. A **while** or **repeat** statement is often chosen when the number of times the loop is to be repeated is determined within the loop and is not known on entry.

Consider the following problem. Suppose you want to compute the average of several numbers. The following program requests that the user first establish the number of numbers to be averaged and subsequently goes through a loop to enter the numbers. Once all numbers have been entered, the average is computed and printed. Type this program into the computer and use it to determine the average weight of the last 10 people you met, your average grade in some course to date, or the average points scored per game for the Dallas Cowboys' football team.

```
program AverageASetOfNumbers;
var
  Total, Index                    :Integer;
  Sum, Number, Average            :Real;
begin
  write('Total of numbers to be entered: ');
  readln(Total);
  Sum := 0;
  for Index := 1 to Total do
    begin
      write('Enter number : ');
      readln(Number);
      Sum := Sum + Number
    end;
  (****Compare and print****)
  Average := Sum / Total;
  writeln('The average of ', Total, ' numbers is ', Average:7:2);
end.
```

In this program you knew in advance the number of numbers that were to be entered. Suppose the data were organized such that the number of numbers was not readily available at the time of input. This could occur if there were too many numbers to count easily. An alternate approach to the averaging problem is to use a special terminator at the end of the data and to continue reading and summing numbers until this terminator is found.

In the following program it is assumed that the last number in the data is followed by a negative number. This implies that negative numbers may not form part of the data themselves and that you may interpret this negative value as the terminator.

```
program AnotherAverageASetOfNumbers;
var
  Count                          :Integer;
  Sum, Number, Average           :Real;
begin
  Sum := 0;
  Count := 0;
  write('Enter number to be averaged: ');
  readln(Number);
  while Number >= 0 do
    begin
      Sum := Sum + Number;
      Count := Count + 1;
      write('Enter number to be averaged: ');
      readln(Number);
    end; (****of while-loop****)
  Average := Sum / Count;
  writeln('The average of ', Count, ' numbers is ', Average:6:2);
end.
```

A **while** loop was used in this program because the number of times it is to be obeyed is not known at the outset. A **while** statement is preferred on the grounds that the terminator may possibly occur as the first number, and this contingency cannot be allowed for naturally using a **repeat** statement with the conditional test at the end of the loop.

12.6 SUBPROGRAMS AND PROCEDURES

Subprograms are self-contained sections of a program which can be included in a complete program. They are building blocks for constructing programs. We can assemble a program from subprograms, rather than directly from Pascal statements, just as a sailing ship is assembled from parts such as hull, mast, and sails, rather than directly from fiberglass, wood, and canvas.

Using such building blocks offers three advantages.

- When you are working on one building block, you can focus your attention on that part of the problem alone, allowing you to break your work down into manageable parts.
- In a large programming project, different people can work on different building blocks at the same time.
- If a building block is needed more than one place in the program, you can write it once and use it as many times as needed. The use of building blocks is a powerful and useful concept.

In Pascal, there are two types of subprograms that can be called on to perform special computational subtasks. They are called **procedures** and **functions**.

A program might contain many procedures. One and only one portion of the program controls the sequence of steps of the computation. This portion of the program is called the *main program*. The main program does some, but not necessarily all, of the computation. It calls on procedures for assistance. Each procedure carries out a special computational subtask and can be called on by the main program as many times as the main program desires. In general, when the main program calls on a procedure, it supplies the needed data, and the procedure returns to the main program with the desired results. A procedure does not initiate any computation on its own; it is totally inactive until called on by the main program.

The advantage to designing a program as a set of procedures which carry out independent tasks is that it makes the program easier to write and understand. The main program can consist of a sequence of calls to procedures to carry out their tasks, so you can see from the main program what the successive tasks are.

Calling And Defining A Procedure

Procedure is the name given to a self-contained section of a program which can be included in a complete program. Making up and using a procedure in Pascal takes place in two parts:

- A procedure declaration, in which a new procedure is defined
- One or more procedure calls that cause the procedure to be obeyed

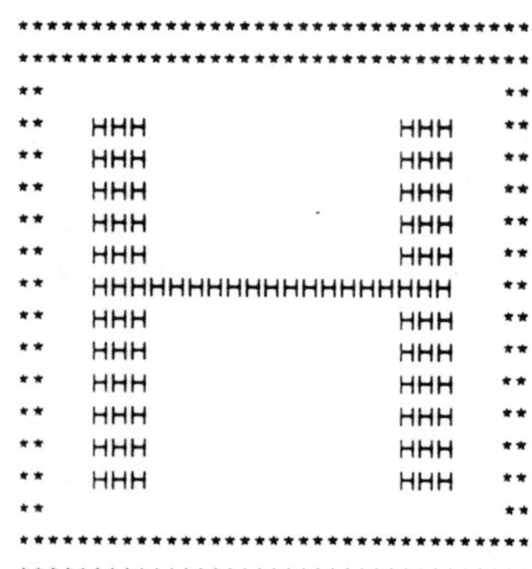

Personal logo "H"

In this way, a clear distinction is drawn between defining the action of a procedure and actually executing the statements in the definition. An example will help to make this distinction clear.

Suppose you want to place a distinctive personal logo at the head of output from your programs. Using a sequence of writeln statements containing character strings, you can have the computer print an initial letter enclosed in a rectangle, such as the one shown in the margin. Rather than place the writeln statements directly into the program after the first **begin**, you may declare a **procedure** at the top of the program as follows:

```
program Logo;
  (****var declarations for the main program****)
procedure Logoprint;
begin
  writeln('************************************');
  writeln('************************************');
  writeln('**                                **');
  writeln('**    HHH                  HHH    **');
  writeln('**    HHH                  HHH    **');
  writeln('**    HHH                  HHH    **');
  writeln('**    HHH                  HHH    **');
  writeln('**    HHH                  HHH    **');
  writeln('**    HHHHHHHHHHHHHHHHHHHHHHHH    **');
  writeln('**    HHH                  HHH    **');
  writeln('**    HHH                  HHH    **');
  writeln('**    HHH                  HHH    **');
  writeln('**    HHH                  HHH    **');
  writeln('**    HHH                  HHH    **');
  writeln('**    HHH                  HHH    **');
  writeln('**                                **');
  writeln('************************************');
  writeln('************************************');
end; (****of procedure Logoprint****)
begin
  (****start of main program****)
  Logoprint;
  (****other program statements****)
end. (****of program Logo****)
```

Note that a procedure declaration always starts with the word **procedure** followed by an identifier chosen by the programmer. The actual statements to be obeyed when the procedure is called are enclosed by **begin** and **end.** Thus, the action of printing the logo is initiated by simply writing the name of the procedure as a statement in the program itself. This procedure call has the same effect as if the writeln statements were written in place of Logoprint at this point.

Because the previous use of a procedure has lengthened the program, you may want to look at what has been achieved:

- At any later state in the program, you can print a further copy of the logo using a single short statement.
- The declaration of Logoprint is clearly distinct from the main program statements and can be taken out and used in other programs very easily.
- By removing the detailed part of printing a logo to a procedure declaration, the main program is rendered less cluttered and hence more readable.

In many applications of procedures it is the third point that is the most important. Procedures can be used to separate single, easily identifiable tasks from the body of a program and enhance the ease with which that program can be written and understood. It is also easier to test and remove errors from a large program if it is made up of logically distinct units.

A procedure declaration serves to define a procedure within the current program and is activated from a procedure statement. Upon completion, program execution continues with the statement immediately following the calling statement. Procedure declarations are written before the **begin** of the main program but after any **var** declarations. Any number of procedure declarations can be contained in a program; each declaration consists of a procedure heading and a procedure body separated by a semicolon. The essential role of the procedure heading is to associate a name or identifier with the procedure being declared. The action to be carried out each time the procedure is called is defined by the procedure body.

A procedure statement serves to activate a previously defined procedure. The statement consists of a procedure identifier, optionally followed by a parameter list (a list of variables or expressions separated by commas and enclosed in parentheses). When the procedure statement is encountered during program execution, control is transferred to the named procedure, and the value of the possible parameters are transferred to the procedure. When the procedure finishes, program execution continues from the statement following the procedure statement. Examples of procedure statements are

```
Find(Name, Address);
Sort (Address);
Area(Radius, Height);
```

Parameters

Values may be passed to subprograms through **parameters.** Parameters provide a substitution mechanism which allows the logic of the subprogram to be used with different initial values, thus producing different results.

The procedure statement which invokes the procedure may contain a list of parameters, called the **actual parameters**. These are passed to the **formal parameters** specified in the subprogram heading. The order of parameter passing is the order of appearance in the parameter lists. Pascal supports two different methods of parameter passing: by *value* and by *reference*, which determines the effect that changes of the formal parameters have on the actual parameters.

Value Parameters When parameters are passed by value, the formal parameters have no effect on the actual parameter. The actual parameter may be any expression, including a variable, with the same type as the corresponding formal parameter. Such parameters are called **value parameters** and are declared in the procedure heading in the following manner:

```
procedure Example(Num1, Num2, Num3 : Integer);
```

Num1, Num2, and Num3 are formal parameters to which the value of the actual parameters are passed. The types of formal and actual parameters must correspond.

A value parameter is used when the parameter's only role is to carry a value into a procedure. The formal parameter simply denotes a value which is supplied by the calling statement; therefore, the corresponding actual parameter may be any expression which produces a value of the required type.

```
procedure SurfaceArea(Radius,Height:Integer);
(****Computes the surface area of a cylinder****)
var
  Area      :Real;
begin
  Area := 2.0 * Pi * (Radius * (Radius + Height);
  writeln('Surface area of cylinder is ' Area:7:2);
end.
```

The previous procedure accepts two integer values representing the radius and height of a cylinder and computes and writes out the surface area of the cylinder. In this case Radius and Height are value parameters. (The SQR function is discussed on page 173.) The corresponding actual parameters in statements calling the procedure may be any expressions producing integer values. For example

```
SurfaceArea(12,8)
SurfaceArea(Rad,Ht)
SurfaceArea(SQR(A),10)
```

are all acceptable calling statements for this procedure, assuming Rad, Ht, and A are integer variables.

Variable Parameters When a parameter is passed by reference, the formal parameter represents the actual parameter throughout the execution of the subprogram. Any changes made to the formal parameter are thus made to the actual parameter, which must therefore be a variable. Parameters passed by reference are called variable parameters, and are declared as follows:

```
procedure Example (var Num1, Num2, Num3 : Integer);
```

The following program uses a procedure to swap two integer values.

```
program SwapExample;
var
  Orange, Plum, Apple, Banana, Pear, Cherry    :Integer;
procedure Swap(var V1, V2 : Integer);
var
  Temp     :Integer;
begin (****procedure Swap****)
  Temp := V1;
  V1 := V2;
  V2 := Temp
end; (****of procedure Swap****)
begin (****main body of program SwapExample****)
  Orange := 24;
  Plum := 24;
  Apple := 18;
  Banana := 28;
  Pear := 42;
  Cherry := 39;
  Swap(Orange, Plum);
  Swap(Apple, Cherry);
  Swap(Pear, Banana);
end. (****of program SwapExample****)
```

The integer variables V1 and V2 are parameters of the procedure Swap. Each time Swap is called, the actual variables in the call (say Orange and Plum) are used in place of V1 and V2. Because of this, V1 and V2 are called formal (or dummy) parameters, while Orange and Plum are known as actual parameters.

As shown in the following example, variable and value parameters may be mixed in the same procedure:

```
procedure Example (var Length, Width : Integer; Num1 : Real);
```

Length and Width are variable parameters and Num1 is a value parameter.

Local And Global Variables

Variables declared in a subprogram are said to be **local** to a subprogram. Variables not declared in a subprogram, but accessible from it, are said to be **global** to the subprogram. In most cases global variables are declared in the main program.

The statements in a subprogram can refer to a global variable provided there is no local variable having the same name. They can use the values of the global variables and assign new values to them. Thus, global variables provide another way of passing data to and from a procedure or function. Values to be passed to a procedure or function can be assigned to global variables before the procedure or function is called. The procedure or function can assign values to global variables, and these values can be accessed by the main program after the procedure or function returns.

If a variable is required only locally within a procedure, it is always best to declare it as a local variable. More often than not, however, a procedure will be required to take in one or more values from the main program and to return one or more values on completion of a call; therefore, some form of communication of values between a program at the time of calling the procedure and the procedure itself is required. The following example uses a procedure to exchange the values in the variables Red and Black.

```
program Exchange;
var
  Red, Black      :Integer;
procedure Switch;
var
  Temp      :Integer;
begin (****procedure Switch****)
  Temp := Red;
  Red := Black;
  Black := Temp;
end; (****procedure Switch****)
begin (****main program****)
  (************************************)
  (************************************)
  (*****statements in the main program****)
  (************************************)
  (************************************)
  Switch;
  (************************************)
  (*****statements in the main program****)
  (************************************)
end. (****main program****)
```

In this program, the exchange sequence within the procedure refers directly to the global variables Red and Black, thus achieving the necessary exchange of values between these variables. Hence, global variables are perhaps the easiest means of communicating values to and from a procedure.

In order to make a procedure more independent of the main program, it is usually preferred to use formal parameters instead of global variables. For example, you can redefine the procedure declaration for Switch to accept the two variables as parameters.

```
procedure Switch (var First, Second : Integer);
var
  Temp      :Integer;
begin
  Temp := First;
  First := Second;
  Second := Temp;
end;
```

The two parameters First and Second exist uniquely in the procedure declaration to mark the places that will be filled at a procedure call by the actual variables whose values are to be exchanged. When two variables are to exchange values in the program, you write a procedure call which includes the variables as parameters; for example,

Switch(Red, Black)

Red and Black are the actual parameters. The action on the calling procedure is effectively: Replace every occurrence of the formal parameter First in the procedure by the actual parameter Red, and every occurrence of Second by Black; then obey the resulting statements in the procedure. On completion of the call of the procedure, the original values of Red and Black will exchange. You may subsequently call the same procedure with different parameters; for example,

Switch(Green, Yellow)

This process achieves an exchange of values between the actual parameters in this call, which are Green and Yellow, assuming that Green and Yellow are added to the integer declarations at the top of the program.

12.7 FUNCTIONS

A procedure is used to identify a set of actions. A **function** identifies a calculation or expression by associating a name and parameters with the calculation of a value. In Pascal there are two types of functions: *standard functions* are predefined and included as part of the Pascal system; and *user-defined functions* are defined by the programmer.

Standard Functions

Pascal has a number of **standard functions** that you can use in programs without having to define them. These functions may be used to perform some commonly required tasks such as finding trigonometrical values or the square of a number.

To see how a function works, let's look at a simple example, the function SQRT. SQRT computes the square root of a number. The square root of a number is the number which, when multiplied by itself, gives the original number. For example, the square root of 9 is 3, and the square root of 25 is 5.

To use a function, you write a function designator, consisting of the name of the function followed by the value to be operated on (the value enclosed in parentheses). Thus

SQRT(49)

means that SQRT is to be applied to the value 49. The value in parentheses is called the argument or parameter of the function, and the function designator represents the value that results when the function is applied to the parameter. Thus, SQRT(49) represents 7, the square root of 49. Here are other examples:

SQRT Function	Value
SQRT(81)	9
SQRT(196)	14
SQRT(6400)	80

The actual parameter of a function can itself be an expression. In that case the value of the actual parameter is worked out before applying the expression. Here are some examples:

SQRT Function	Value
SQRT(20+5)	5
SQRT(25*4+21)	11
SQRT(600/5+24)	12

Function designators can be used as part of larger expressions. The values of the function designators are worked out before any of the other operators are applied. For example:

Expression	Value
SQRT(144)+13	25
5*SQRT(625)	30
SQRT(100/5+5)+SQRT(400)	45

Pascal has a number of standard functions. Some of the most useful ones are defined here.

ABS(X) Computes the absolute value of X. The parameter X must be either real or integer, and the result is the same type as the parameter. If the parameter is negative, the corresponding positive value is returned; if the parameter is positive, its value is returned unchanged. Thus, the value of both ABS(-84) and ABS(84) is 84.

INT(X) Returns the integer part of X (i.e., the greatest integer number less than or equal to X, if X >= 0, or the smallest integer number greater than or equal to X, if X < 0. The parameter X must be either real or integer and the result is real. Thus, the value of INT(41.7) and INT(-7.8) is 41.0 and -8.0, respectively.

SQR(X) Computes the square of the parameter X. The value of SQR(9) is 81, and the value of SQR(2.5) is 6.25. The parameter X must be either real or integer, and the result is the same type as the parameter.

SQRT(X) Computes the square root of the parameter X. The square root of a number is the number which, when multiplied by itself, gives the original number. For example, the square root of 25 is 5.0. Thus, the value of SQRT(144) is 12.0, and the value of SQRT(6.25) is 2.5. The parameter X must be either real or integer, and the result is real.

TRUNC(X) Converts a real number to an integer by discarding everything to the right of the decimal point. Thus, the value of both TRUNC(63.3) and TRUNC(63.9) is 63.

ROUND(X) Converts a real number to an integer by rounding the real value to the nearest integer. Thus, the value of both

Pascal also includes as standard functions several trigonometric and logarithmic functions:

SIN(X)	sine of X in radians
COS(X)	cosine of X in radians
EXP(X)	e(2.71828) to the power X
LN(X)	natural logarithm of X
LOG(X)	logarithm base 10 of X
ATAN(X)	arc tangent of X in radians

x^y := EXP(y * LN(x))

Pascal does not include an operator or function for raising an arbitrary number to an arbritrary power. Since such numbers are often calculated through the use of logarithms, the designer of Pascal decided to leave this function for the user to program by simply taking the logs and reexponentiation using the EXP function. For example,

$2^{18.4}$ is calculated by

TWO18 := EXP(18.4 * LN(2)).

ROUND(4.5) and ROUND(4.77) is 5, and the value of ROUND(4.2) is 4.

RANDOM Generates a random number.

Two of the standard functions, TRUNC and ROUND, are called *transfer functions* because a real argument is converted to an integer result in each case. Consider an assignment to determine the approximate equivalent value in degrees Celsius of a temperature in degrees Fahrenheit. This is accomplished by using the formula

DegreeCelsius := 5/9 * (DegreeFahrenheit - 32)

If you wanted to round the result to the nearest integer value you can use

DegreeCelsius := ROUND(5/9 * (DegreeFahrenheit - 32))

Alternatively, if you insist on a value in degrees Celsius that is on no account greater than the real result, the result is given by

DegreeCelsius := TRUNC(5/9 * (DegreeFahrenheit - 32))

Sample Program — Area Of A Triangle

Heron's formula may be used to find the area of any triangle, given the measurements of the three sides. The formula is

$$\text{Area} = \sqrt{s(s - a)(s - b)(s - c)}$$

c

b

a

where s = ½(a + b + c) and a, b, and c are sides of the triangle. The following program computes the area of a triangle using Heron's formula.

```
program HeronsFormula;
var
   Area, S, A, B, C       :Real;
begin
   write('Enter side A: ');
   readln(A);
   write('Enter side B: ');
   readln(B);
   write('Enter side C: ');
   readln(C);
   S := (A + B + C) / 2;
   Area := SQRT(S * (S - A) * (S - B) * (S - C));
   writeln('Triangle area is ', Area:6:2);
end.
```

HERON OF ALEXANDRIA

Heron (or Hero) was a Greek mathematician and inventor who lived some time between the 2nd century B.C. and the 3rd century A.D. He established a formula for the area of a triangle in terms of its sides, invented a steam engine, and other devices powered by water, steam or compressed air.

Sample Program — Volume

The following program computes the volume of a cylinder using the formula

$$\text{Volume} = \pi r^2 h$$

where r is the radius, and h is the height. The Greek letter π represents the mathematical constant Pi whose approximate value is 3.1415926. . .

To translate this formula into Pascal, use the variables Volume, Radius, and Height and the SQR function to square the radius.

```
program Volume;
  (****Computes volume of cylindrical container****)
const
  Pi = 3.14159;
var
  Radius, Height, Vol      :Real;
begin
  write('Enter Radius: ');
  readln(Radius);
  write('Enter Height: ');
  readln(Height);
  Vol := Pi * SQR(Radius) * Height;
  writeln('Volume of cylinder is ', Vol:8:2);
end.
```

User-Defined Functions

You have seen how to use standard functions such as SQRT and TRUNC. Now you will see how to define our own functions. A function identifies a calculation or expression by associating a name and parameters with the calculation of a value.

The **user-defined function** is defined by the word "function" followed by the name of the function, and a bracketed list of formal parameters. The type of function result must be specified in the heading as well. The general form is

function name (parameter list) : result type

If no parameter list is specified, then no actual parameters can be passed to the function.

Suppose you wanted to perform exponentiation in a program. Pascal's facilities for exponentiation are limited to the squaring function, SQR. You might have a need, however, for exponentiation involving powers greater than or less than 2. What you really want is a function that lets you evaluate expressions such as

$$8.0^{4}$$
$$2.0^{-2}$$

You might then decide to declare a function called POWER. This

DID YOU KNOW?

The LOG(X) can always be obtained by dividing LN(X) by 2.303.

function would take two parameters, a real value and an integer value, such as

$$POWER(8.0,4) = 8.0^4 = 4096.0$$
$$POWER(2.0,-2) = 2.0^{-2} = .25$$

Once declared, the following function declaration for POWER can be used just like a predefined standard function.

```
function POWER(Factor:Real; Exponent:Integer): Real;
  (*Function raises a real number to an integer power*)
var
  TempFactor      :Real;
  Count           :Integer;
begin
  if Exponent = 0
    then POWER := 1
    else
      begin
        TempFactor := Factor;
        for Count := 2 to ABS(Exponent) do
          TempFactor := TempFactor * Factor;
        if Exponent < 0
          then POWER := 1 / TempFactor
          else POWER := TempFactor;
      end;
end; (****of POWER function****)
```

The first line of the declaration is the function header. The function header specifies the name of the function, the number and type of its formal parameters, and the type of the value returned by the function. The function POWER has two formal parameters: Factor (of type Real) and Exponent (of type Integer). Every time the POWER function is called, Factor and Exponent are given the values of the corresponding actual parameters. Within the function body, Factor and Exponent can then be used like ordinary variables.

Following the header, the next element of a function declaration is its internal declarations. Function POWER has internal declarations for two variables: TempFactor and Count. Variables declared internally (local variables) are not available to the main program; they are temporary variables to be used only within the function. TempFactor is used inside POWER to accumulate the multiplications; Count is used as a loop counter. After POWER is completed, TempFactor and Count will cease to exist until the next time the function POWER is called.

The final element of a function declaration is the function body itself. The function body comprises the statements that generate the value returned by the function. Any valid Pascal statement can appear in the body of a function. To indicate the value that the function will return, one writes an assignment statement in which the left-hand part is the function name. For instance, the following assignment statements inside POWER are used to assign a resulting value to POWER.

```
POWER := 1
POWER := 1 / TempFactor
POWER := TempFactor
```

Consider another function. The function GASCOST calculates the cost of gasoline for a trip. Dividing the total mileage (TotalMile) by the automobile's miles-per-gallon average (Average) tells us how much gas will be used, and multiplying this value by the price per gallon (Price) gives us the total gasoline cost (GASCOST).

$$\text{GASCOST} = \left(\frac{\text{TotalMile}}{\text{Average}} \right) \text{Price}$$

The following program determines a 2430 mile trip requires $86.79 for gasoline if the car averages 21 miles per gallon, and gasoline costs $.75 per gallon.

```
program GasolineComputation;
var
  TotalMile, Average, Price     :Real;
function GASCOST(TotalMile, Average, Price:Real):Real;
begin
  GASCOST := TotalMile / Average * Price
end; (****of GASCOST function****)
begin
  write('Enter total miles, automobile's average, gas price: ');
  readln(TotalMile, Average, Price);
  writeln('Trip cost is $ ', GASCOST(TotalMile,Average,Price):6:2);
end.
```

```
COMPILE/RUN

Enter total miles, automobile's average,gas price:
  2430        21        0.75
Trip cost is $86.79
```

12.8 RECURSION

Recursion is one of the powerful features of Pascal. Recursion refers to the ability of defining an object in terms of itself. Procedures and functions in Pascal can be recursive — they can call themselves. Here is an example of a recursive procedure.

```
procedure Increment(Number:Integer):Integer;
begin
  Number := Number + 1;
  if Number <= 100
    then Increment(Number);
  writeln(Number:3);
end;
```

In this example, procedure Increment continues calling itself until the value of Number exceeds 100. Recursion is useful when there is a set of steps that must be repeatedly executed until a particular condition is met.

As another example, consider a function that computes the greatest common divisor (GCD) of two integers (the GCD of two integers is the greatest integer that is a factor of each of the given integers).

```
function GCD(M,N:Integer):Integer;
begin
  if N = 0
    then GCD := M
    else GCD := GCD(N, M mod N);
end;
```

Consider an additional example. The factorial of the value of N is the product of all the integers from 1 through the value of N. The function is defined as

$$N\text{-factorial} = N! = N \times (N - 1) \times (N - 2) \ . \ . \ . \ 1$$

where N > 0. N! = 1 when N = 0. For example,

$$1! = 1$$
$$2! = 2 \times 1 = 2 \times 1! = 2$$
$$3! = 3 \times 2 \times 1 = 3 \times 2! = 6$$
$$4! = 4 \times 3 \times 2 \times 1 = 4 \times 3! = 24$$
$$5! = 5 \times 4 \times 3 \times 2 \times 1 = 5 \times 4! = 120$$
$$6! = 6 \times 5 \times 4 \times 3 \times 2 \times 1 = 6 \times 5! = 720$$

The following program is the classical demonstration of the use of a recursive function to calculate the factorial of an integer number.

```
program Nfactorial;
var
  Number      :Integer;
function FACTORIAL(N:Integer):Real;
begin (****FACTORIAL function****)
  if N = 0
    then FACTORIAL := 1
    else FACTORIAL := N * FACTORIAL(N - 1)
end; (****of FACTORIAL function****)
begin (****main program****)
  write('Enter N; ');
  readln(Number);
  writeln(Number, '! = ', FACTORIAL(Number):6:0);
end. (****of program Nfactorial****)
```

```
COMPILE/RUN

Enter N; 7
7! = 5040
```

Note that the identifier FACTORIAL is used in two different ways. When FACTORIAL appears on the left-hand side of the assignment operator it refers to the computer's memory location used to store the result returned by the function. When FACTORIAL appears on the right-hand side of the assignment operator it refers to the FACTORIAL function and causes a recursive call to that function.

12.9 ARRAYS

In Chapter 5, you learned about the predefined data types — integer, real, char, and Boolean. A variable of one of these types can hold only one value at a time. For example, if you define

```
var
    Number      :Integer;
```

then the variable Number has only one specific value at any moment. However, there are situations where you would like to have a list of values, such as a list of numbers or characters.

When a computer application involves several items of data that are similar in type and relate to the same basic algorithmic process, it is frequently cumbersome or literally impossible to assign each item a unique variable name. Typical examples are: a list of names, dates, part numbers, or part names for a conventional information processing application; the parameters describing the path of a space vehicle's trip to a far away planet; a collection of observations recorded as numeric data for a statistical analysis; the chemistry test scores of a student; the number pattern that makes up a magic square; and product identification numbers, product names, and prices of items in a supermarket. In cases such as these, the data items are grouped and given a descriptive identifier, which is used to refer to the entire collection of data. Individual data items are referenced through their relative position in the collection. This is where arrays are used.

Definition Of Array

An **array** is a structured type consisting of a fixed number of components which are all of the same type. Each component can be explicitly accessed by indices into the array. *Indices* are expressions of any scalar type placed in square brackets suffixed to the array identifier. The indices are called **subscripts.** The array identifier with subscripts is called a **subscripted variable.**

The following declaration defines an array:

```
Var
    Food      :array[1 . . 500] of Integer;
```

In this example, the name of the array is Food. Food is the array identifier, and an array identifier is selected and used primarily in the same manner that a variable name is selected and used. Each item in the array in the above declaration has an integer value, and there are 500 of these items. Each data item in this array is an integer value because the declaration explicitly specifies integer values with the clause "**of** Integer". You know that this array has 500 data items because the subscript, denoted by brackets runs from 1

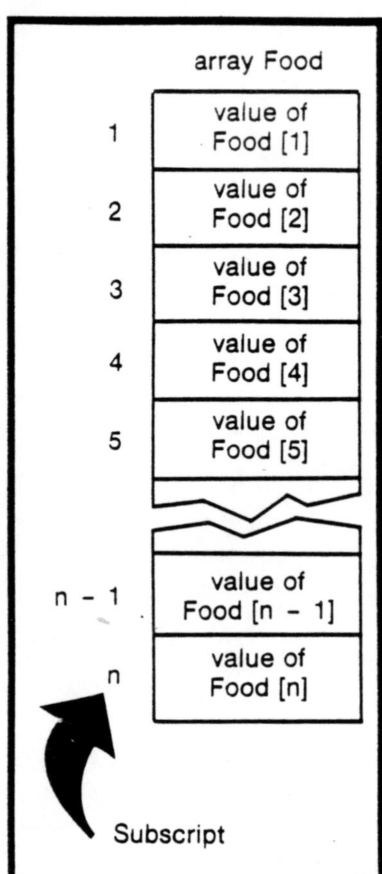

Conceptual view of an array.

to 500. If the subscript had run from 1 to 200, then the array would contain 200 data items.

The subscript denotes a quantity used to select an element of the array. Each entry in an array is called an *element*. Consider the conceptual view of array Food with *n* elements shown in the margin. Each value is referenced with a subscript that denotes its relative position. A subscript of 1 is used to select the first element represented by Food[1]. Similarly, the second element is referenced by Food[2], the third element by Food[3], and so on.

The definition of an array consists of the reserved word **array** followed by: the subscript, enclosed in square brackets; the reserved word **of**; and the element type. Some typical array declarations are

```
Number : array[1 . . 40] of Real;
Date : array[1 . . 31] of Interest;
Month : array[1 . . 12] of Integer;
EmployeeNumber : array[1 . . 300] of Integer;
Set : array[1 . . 8] of Boolean;
  Message : array[a . . z] of Char;
```

One-dimensional Arrays

A **one-dimensional array** is a list or set of storage locations or variables that are the same type, share the same name, but have different subscripts. Consider this information:

Variable Name	Content
PriceOfCamera	210.80
PriceOfFilm	5.20
PriceOfTripod	65.00
PriceOfCase	32.00
PriceOfFilter	8.00
PriceOfFlash	99.55
PriceOfLens	102.50
PriceOfProjector	180.00
PriceOfFrame	12.88

This data could be stored in a one-dimensional array in this manner:

Variable name	Content
Price[1]	210.80
Price[2]	5.20
Price[3]	65.00
Price[4]	32.00
Price[5]	8.00
Price[6]	99.55
Price[7]	102.50
Price[8]	180.00
Price[9]	12.88

The numbers in the square brackets are the distinguishing subscripts. The subscripted variables that make up the array are called *elements*. These elements can be manipulated just like a simple variable, such as

```
read(Price[1], Price[2];
Difference := Price[1] - Price[2];
writeln('Price of Item 2 is ', Price[2];
```

Consider a list of eight numbers and suppose the name of the list is X. Write X[1] to refer to the first element in the list, X[2] to refer to the second element in the list, and X[8] to refer to the last element in the list. You can find the sum of all eight numbers in the list through the statement

```
Sum := X[1] + X[2] + X[3] + X[4] + X[5] + X[6] + X[7] + X[8]
```

This statement would accomplish the task, but what if the list contained 200 numbers and you wanted to add all 200? Obviously, you must find a more efficient way. The following program segment uses a program loop to add 200 elements of a list named X. The sum of the elements would now be stored in variable Sum.

```
Sum := 0;
for Index = 1 to 200 do
  Sum := Sum + X[Index];
```

The following program reads 10 test scores for a particular student, averages those test scores, and prints the result.

```
program GradeAverage;
var
  Test          :array[1 . . 10] of Integer;
  Average       :Real;
  Sum, Index    :Integer;
begin
  (****Set summation variable to zero****)
  Sum := 0;
  (****Read test scores into array Test and****)
  (****obtain sum of test scores****)
  for Index := 1 to 10 do
    begin
      write('Enter score for Test # ', Index, ' : ');
      readln(Test[Index]);
      Sum := Sum + Test[Index];
    end;
  (****Compute average of test scores****)
  Average := Sum / 10;
  writeln('Average grade is ', Average:4:1);
end.
```

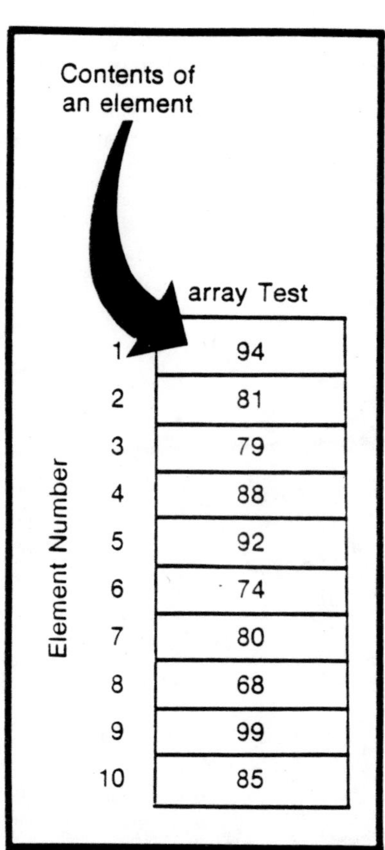

One-dimensional array containing a student's 10 test scores.

The student's 10 test scores are stored in an array called Test. Pictorially the array can be visualized as shown in the margin. The identifier Test refers to the entire array. Each element of the array contains a value. These elements are numbered from 1 to 10. Note that the element number is distinctly different from the contents of that element.

	Column 1	Column 2	Column 3	Column 4	Column 5
Row 1	1	0	0	1	0
Row 2	0	1	(1)	0	1
Row 3	1	1	1	0	0
Row 4	0	1	0	1	0
Row 5	1	0	0	1	1

Array Number

In the program, statements in the **for**-loop cause the 10 test scores to be stored in array Test and their sum to be accumulated in variable Sum. The program then finds the average by dividing the sum of the test scores by 10. The writeln statement causes the average to be printed.

Two-dimensional Arrays

A **two-dimensional array** is composed of horizontal rows and vertical columns. To find any element in a two-dimensional array, you must identify two subscripts: (a) one that indicates the row (row subscript); and (b) one that indicates the column (column subscript). The subscripts are appended to the array identifier within brackets. Suppose that Number is the name of a two-dimensional array with 5 rows and 5 columns. Then Number[2,3] would refer to the second element in the third column (see figure in margin).

Many techniques in science, engineering, and commerce deal with data which are organized in two dimensions. Pictures from interplanetary explorers, for example, are enhanced and analyzed by computer. The pictures are represented inside the computer as a two-dimensional array of numbers. Each number corresponds to the brightness of a picture element or point. A picture is converted into a table of numbers by a special input device, and after it is processed, it is converted back into a picture by a special output device. It is easier and more natural for a programmer to think in terms of a two-dimensional set of picture elements — the picture retaining its two-dimensional form when referred to in the program — than it would be if the picture elements were strung out row-wise or column-wise into a one-dimensional array or list.

I've had it with hunting and gathering. I think I'll become a programmer!

12.10 STRINGS

Pascal offers the convenience of string types for processing character strings (i.e., sequences of characters). String types are structured types, and in many ways they are similiar to array types. Strings with a constant length may be thought of as character arrays. Character arrays are arrays with one subscript and components of the type char, such as

array[1 . . n] **of** Char

where *n* equals the number of characters in the string.

You have already encountered strings as parameters to write and writeln. The following writeln statement prints a string on the user's terminal:

```
writeln('Programming computers with Pascal')
```

This string contains 33 characters, and it is logical to think of this string as an array of characters.

The definition of a string type must specify the maximum number of characters it can contain. The definition consists of the reserved word **string** followed by the maximum length enclosed in square brackets. The length is specified by an integer constant in the range 1 through 255.

```
var
  State = string[20];
  Name = string[18];
  LongString = string[255];
```

The variable State could hold up to 20 characters; Name could only hold up to 18 characters, therefore, the statement

```
Name := 'Jefferson Williamson';
```

would only store the first 18 characters into Name. The last variable, LongString, represents the maximum length possible for a string — 255 characters.

String Expressions

Strings are manipulated by the use of **string expressions**, which consist of string constants, string variables, function designators, and operators.

The plus sign may be used to link strings together. For example,

```
'ABC' + 'XYZ'                       = ABCXYZ
APPLE  ' + 'PASCAL'                 = APPLE PASCAL
'Windsurfing  ' + 'is  ' + 'fun'    = Windsurfing is fun
'84' + '.' + '22'                   = 84.22
'X ' + 'Y' + ' Z'                   = X Y Z
```

The relational operators =, <> , < ,<=, > , and >= are lower in precedence than the concatenation operator(+). When applied to string operands, the result is a Boolean value (True or False). When comparing two strings, single characters are compared from the left to the right according to their value. Strings are equal only if their lengths as well as their contents are identical. For example,

'CHINA' = 'CHINA '	is False
'FRANCE' = 'FRANCE'	is True
'X' < 'Y'	is True
'R' > 'r'	is False
'DON' <> 'DAN'	is True
'SHERRIE' < 'SHERRY'	is True
'Washington' > 'Nevada'	is True

		48	**0**
		49	**1**
		50	**2**
		51	**3**
		52	**4**
		53	**5**
		54	**6**
		55	**7**
		56	**8**
		57	**9**
65	**A**	97	**a**
66	**B**	98	**b**
67	**C**	99	**c**
68	**D**	100	**d**
69	**E**	101	**e**
70	**F**	102	**f**
71	**G**	103	**g**
72	**H**	104	**h**
73	**I**	105	**i**
74	**J**	106	**j**
75	**K**	107	**k**
76	**L**	108	**l**
77	**M**	109	**m**
78	**N**	110	**n**
79	**O**	111	**o**
80	**P**	112	**p**
81	**Q**	113	**q**
82	**R**	114	**r**
83	**S**	115	**s**
84	**T**	116	**t**
85	**U**	117	**u**
86	**V**	118	**v**
87	**W**	119	**w**
88	**X**	120	**x**
89	**Y**	121	**y**
90	**Z**	122	**z**

A char value is one character in the ASCII character set (see above). Characters are ordered according to their ASCII value; for example: 'A' is less than 'B', '3' is less than '7', and 'm' is greater than 'd'.

Suppose you are sorting a list of names into alphabetical order. At some point you will compare two strings to find which comes before (is less than) the other. The statement

if Name1 > Name2 **then** . . .

will take some action if and only if Name2 comes before Name1. The comparison results in one of three cases:

- Both strings are the same length and have the same contents. In that case, they are equal.
- Both strings have the same contents up to the end of one string; the other string has additional characters beyond that point. In that case, the shorter string is less than the longer one.
- The strings cease to match at some point (it may even be the first character). In that case, the string whose unmatched character has the lower numeric value is less than the other string.

String Assignments

The assignment operator is used to assign the value of a string expression to a string variable, such as

```
Name := 'Wilson';
Address := '48 Collins Avenue';
Score := '30' + Test;
```

In the last example, if the string variable Test contained the string '80', then after the assignment, the variable Score will contain 3080.

You can also read and write strings by using read, readln, write, and writeln statements. For example, the following program allows you to type in a line (up to 100 characters); then three copies of what you typed appear on the screen. It continues to do this until you type, "I Quit!"

```
program Copy;
var
  I        :Integer;
  Line     :String[100];
begin
  writeln('Entering copy mode — type "I Quit!" to stop');
  repeat
    readln(Line);
    for I := 1 to 3 do
      writeln(Line);
    until Line = 'I Quit!';
end. (****of program Copy****)
```

Strings and Characters String types and the type char are compatible. Thus whenever a string value is expected, a char value may be specified instead, and vice versa. Furthermore, strings and characters may be mixed in expressions.

The characters of a string variable may be accessed individually through string indexing. This is achieved by appending an index (subscript) expression of type Integer, enclosed in square brackets, to the string variable. For example, you can reference individual characters in the string variable Chess using the notation Chess[1], Chess[2], Chess[3], and so on. The first location, Chess[0], contains the current length of Chess. If you execute the statement

```
Chess := 'Move the King';
```

then Chess[0] contains the value 13, since there are 13 characters in 'Move the King'.

12.11 ENUMERATED TYPE

An **enumerated type** has the form (identifier, identifier, identifier, . . .). Each identifier is a newly defined value; it names a member of an enumerated set. The identifiers are values coined by the programmer. For example, (BLUE, RED, WHITE) defines the type whose values are the three colors in the American flag. The order of appearance also implies an ordering of value from lowest to highest. Enumerated types can be assigned and compared but not read or written. For example, if the programmer knows that a particular variable can have only three values, an enumeration may be used to name those values explicitly. For example,

```
var
  HatSize : (Small, Medium, Large);
```

Subsequent program statements could make use of the ordering implied in the enumeration, such as

```
if HatSize > Small then
  for HatSize := Small to Large do
```

If enumeration types were not possible, the programmer would have to resort to declaring HatSize as an integer and assigning numeric

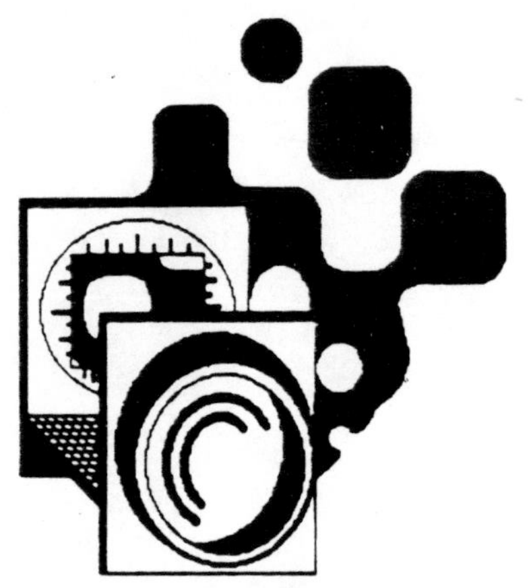

If you continue your study of the Pascal programming language, you will learn instructions that will allow you to draw graphs, diagrams and other pictures.

values to each possible size. Enumeration is a feature of Pascal that allows the programmer to produce meaningful program statements by coining names.

12.12 FURTHER PASCAL

In attempting to introduce the reader to the fundamentals of computer programming it has been necessary to omit exhaustive descriptions of certain Pascal features in order to emphasize general principles. A handful of Pascal topics have not been mentioned at all. Although the omitted material is small in number, this is not meant to imply that the extra features are never important. The following brief notes are included to give some idea of items you have missed.

Sets

In mathematics, a set is a collection or group of objects. For example, here is a set composed of the counting numbers:

$$\{1, 2, 3, 4, 5, 6, 7, \ldots\}$$

Use braces to enclose the objects making up a set. The objects belonging to the set are called the elements, or members, of the set. For example, 3 is an element of the set of counting numbers. The three dots show that the elements continue in the same way but that the others are not actually listed; there is no last number in the set of counting numbers.

In Pascal, sets work the same way. They are enclosed in rectangular brackets rather than braces, however. Here is a set, written Pascal style:

[2, 4, 6, 8, 10, 12, 14, 16, 18, 20]

Sets are collections of data items of the same type. They are similar to arrays, except that there is no subscript. Therefore you cannot access any one individual element of a set the way you could pick an individual element of an array. The only things you can do with sets are to combine them in various ways, to compare them as to their contents, and to test whether or not a data item is an element of the set.

Two sets are equal if and only if their elements are the same. There is no ordering involved, so the sets [1, 2, 3], [3, 2, 1] and [2, 1, 3] are all equal.

There are three operations involving sets, similar to the operations addition, subtraction, and multiplication operations on numbers: The *union* (or sum) of two sets A and B (written A + B) is the set whose elements are in either A or B. For instance, the union of [1, 3, 5, 7] and [2, 3, 4] is [1, 2, 3, 4, 5, 7]. The *intersection* (or product) of two sets A and B (written A * B) is the set whose elements are in both A and B. Thus, the intersection of [1, 3, 4, 5, 7] and [2, 3, 4] is [3, 4]. The *difference* of B with respect to A (written A - B) is the set whose members are members of A but not of B. For instance, [1, 3, 5, 7] - [2, 3, 4] is [1, 5, 7].

Records

An array is a suitable storage structure to use if you have a set of the same type of elements that you wish to consider as a unit. If the elements are different types, or if you wish to distinguish elements by name rather than by an index, a **record** is likely to be more appropriate. The components of a record are accessed by name rather than by subscript.

A record is a group of related items that make up one logical unit of information. For example, in a program for handling a list of names and addresses, a record might be comprised of first name, last name, address, city, state, zip code, and telephone number. Let the record be called PEOPLE. Now, you can access the CITY part of the record by using PEOPLE.CITY. The telephone part of the record can be accessed by PEOPLE.TEL, and so on. Having access to the various parts of the records allows you to manipulate data in a more efficient manner. Suppose you want to sort the name and address records of the following list so that they are in order by last name.

Tod Gardner
Steve Wilson
Rochelle Nelson
Keith Kline
Robert Brown
Selena Allison

Sorted by last name, the order would be

Allison Selena
Brown Robert
Gardner Tod
Kline Keith
Nelson Rochelle
Wilson Steve

The fields of a record may be records themselves and various records may be linked to one another with pointers (variables that hold the location of another variable). Records may have fixed parts and variant parts. The **with statement** may be used to make assignments to fields easier. For example

```
with SUBJECT do
  begin
    Sex := Male;
    Married := True;
    Birthdate := 1962;
    SocialSecurity := 432618402;
  end;
```

Files

Files provide a program with channels through which it can pass data. A file can either be: (a) a disk file, in which case data are written to and read from a magnetic disk or the computer's standard input/output channels; or (b) the keyboard and the visual display screen.

A file consists of a sequence of components of equal type. The number of components in a file is not determined by the definition of the file. Instead, the Pascal system keeps track of the file accesses through a file pointer, and each time a component is written to or read from a file, the file pointer of that file is advanced to the next component. Because all components of a file are of equal length, the position of a specific component can be calculated. Thus, the file pointer can be moved to any component in the file, providing random access to any element of the file.

Space and permanence are the main reasons for using disk files. For most practical purposes, there is no limit to the amount of storage space available on this media. In addition, information can be copied to disk and left there until required again, perhaps by another program.

Pointer Structures

Variables discussed up to now have been static (i.e., their form and size is predetermined), and they exist throughout the entire execution of the block in which they are declared. Programs, however, frequently need the use of a data structure which varies in form and size during execution. Dynamic variables serve this purpose because they are generated as the need arises and may be discarded after use. Dynamic variables allow flexible run-time management of storage.

Dynamic variables are not declared in an explicit variable declaration like static variables, and they cannot be referenced directly by identifiers. Instead, a special variable containing the memory address of the variable is used to point to the variable. This special variable is called a **pointer variable.**

Basically, a variable may be defined as being a pointer to an object of another type and it becomes, effectively, a reference to that other object. Pointers are used almost exclusively as references to records in complex data structures. In its simplest form, a list of records might be created using pointers rather than an array of records. To do this, each record would contain a pointer to the next record in the list so that the elements of the list could be accessed, in order, by moving down the chain of pointers from one record to the next. List elements could be removed from the list by altering the pointers.

DID YOU KNOW?

Any even number *n* can be expressed as the sum of two prime numbers. This conjecture was stated in a letter by the mathematician Goldbach in 1742. It is undoubtedly true, and the number of "Goldbach pairs" (which is three for the even number 22, namely 11 and 11, 5 and 17, 3 and 19) approaches infinity as the even number *n* increases, but an acceptable "mathematical" proof has never been found.

12.13 SAMPLE PASCAL PROGRAMS

$y = x^2 + 5x + 6$

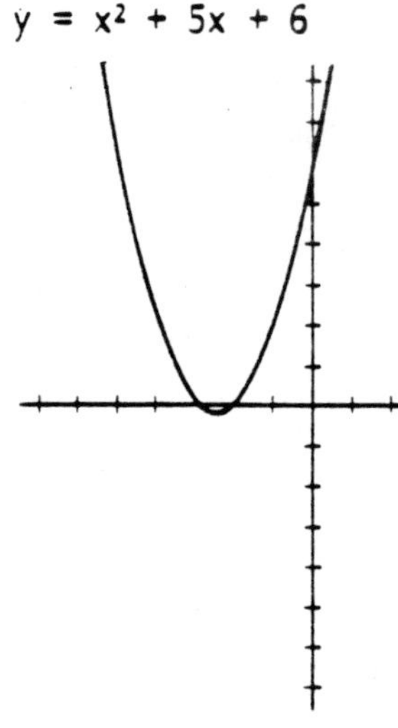

The beginning computer user is advised to start with problems that are relatively easy to understand. With a clear understanding of simple problems one may gradually move on to more complicated problems. The reader should remember that there are often many possible approaches and solutions to any given problem.

Roots Of An Equation

The roots of a quadratic equation of the form:

$$Ax^2 + Bx + C = 0$$

are given by the quadratic formula:

$$\text{root} = (-B \pm \sqrt{B^2 - 4AC})/2A$$

$y = x^2 + x - 2$

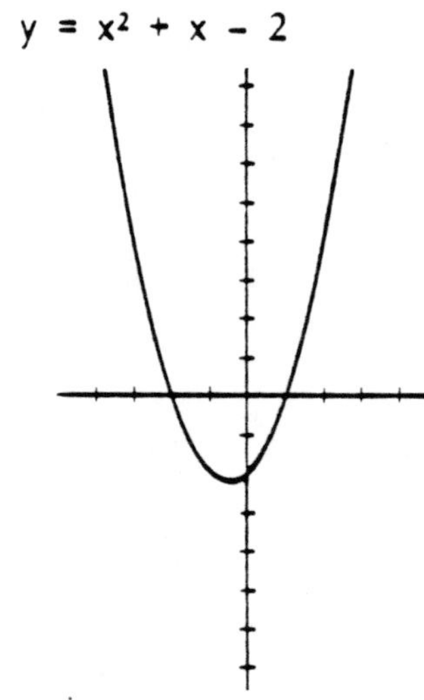

For example, the equation $x^2 + 5x + 6 = 0$ can be factored as $(x + 2)(x + 3) = 0$ and the roots are known as $x = -2$ and $x = -3$. The following program uses the quadratic formula to solve the equations shown in the margin.

```
program RootsOfAnEquation;
var
  A,B,C,Counter        :Integer;
  Root1,Root2          :Real;
begin
  for Counter := 1 to 3 do
    begin
      write('ENTER COEFFICIENTS OF EQUATION: ');
      readln(A,B,C);
      Root1 := (-B + SQRT(B * B - 4 * A * C)) / (2 * A);
      Root2 := (-B - SQRT(B * B - 4 * A * C)) / (2 * A);
      writeln('COEFFICIENTS OF EQUATION ARE: ',A, ' ',B, ' ',C);
      writeln('ROOTS OF EQUATION ARE: ',Root1:6:2,
      writeln;                                ' and ', Root2:6:2);
    end;
end.
```

$y = x^2 + x - 6$

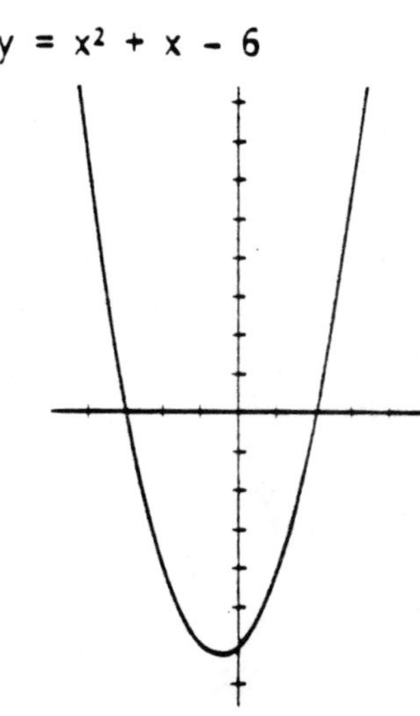

COMPILE/RUN

```
ENTER COEFFICIENTS OF EQUATION: 1, 5, 6
COEFFICIENTS OF EQUATION ARE: 1  5  6
ROOTS OF EQUATION ARE:    -2  -3

ENTER COEFFICIENTS OF EQUATION: 1, 1, -2
COEFFICIENTS OF EQUATION ARE:  1  1  -2
ROOTS OF EQUATION ARE:     1  -2

ENTER COEFFICIENTS OF EQUATION: 1, 1, -6
COEFFICIENTS OF EQUATION ARE:  1  1  -6
ROOTS OF EQUATION ARE:     2  -3
```

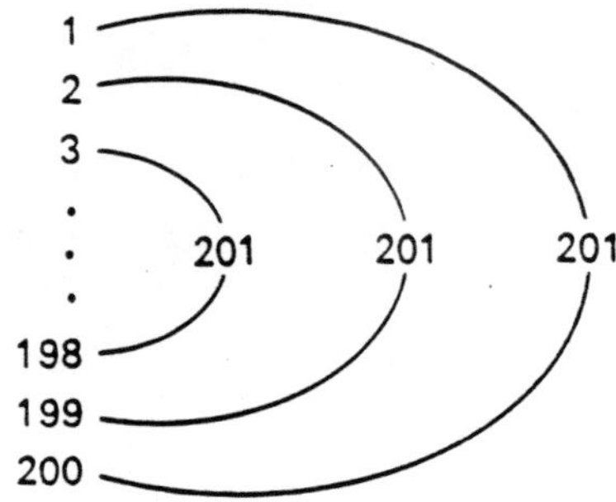

Karl Gauss' Calculation

Have you ever tried to calculate the sum of the first 200 positive integers (1 + 2 + 3 + . . . + 199 + 200)? Karl Gauss, a great German mathematician was able to solve this problem in just a few seconds. He paired off the numbers (1-200, 2-199, 3-198, etc.) knowing that there would be 100 such pairs, each of whose sum was 201. Thus, the sum of the first 200 positive integers would be 100 × 201 = 20100! Was Karl right? The following program proves that Gauss' computational method works.

```
program KarlGaussCalculation;
var
  Sum,GaussSum,Counter        :Integer;
begin
  (**** Sum of the first 200 integers ****)
  Sum := 0;
  for Counter := 1 to 200 do
    Sum := Sum + Counter;
  (**** Integer sum by Karl Gauss method ****)
  GaussSum := 100 * 201;
  writeln('SUM OF THE FIRST 200 INTEGERS = ', Sum);
  writeln('SUM OF GAUSS'' CALCULATION = ',GaussSum);
end.
```

Polynomial

The following program evaluates the 6th degree polynomial

$$Y = x^6 - 3x^5 - 93x^4 + 87x^3 + 1596x^2 - 1380x - 2800$$

for all integer values of x between −12 and +16.

Karl Friedrich Gauss (1777-1855), along with Archimedes and Newton, is considered one of the three greatest mathematicians of all time. While still in grade school, he discovered and proved the formula 1+2+3+...+n=n(n+1)/2. By the time he was 21 he had contributed more to mathematics than most mathematicians do in a lifetime. He is especially famous for his pioneering work in the theory of numbers.

```
program Polynomial;
var
  X, Y       :Real;
function POWER(X:Real;K:Integer):Real;
var
  Result      :Real;
  Index       :Integer;
begin
  Result := 1;
  for Index := 1 to K do
    Result := Result * X;
  POWER := Result;
end; (*POWER function*)
begin
  X := -12.0;
  repeat
    Y := POWER(X,6)-3*POWER(X,5)-93*POWER(X,4)+87*
           POWER(X,3)+1596*POWER(X,2)-1380*X-2800;
    writeln('FOR X =  ', X:2:0, '  ', 'Y =  ', Y:10:2);
    X := X + 1;
  until X > 16;
end.
```

REVIEW EXERCISES

1. The Pascal language was named for ______________ .
2. The Pascal language was developed by ______________ .
3. List the four standard data types found in Pascal.
4. Which of the following are integers.

a. 64	c. -19	e. +53
b. -23.8	d. 263.0	f. 0.03

5. Which of the following are real numbers.

a. 2369	c. -498.0	e. 177.5
b. +42.5	d. 0.019	f. 3.400000E2

6. Represent the number 623.5 in floating-point notation.
7. A series of characters is called a ______________
8. 'FIBONACCI NUMBERS' is an example of a ______________ .
9. A Pascal program consists of a ______________ heading and a program ______________ .
10. Every Pascal program should begin with a ______________ .
11. List the four declaration parts of a Pascal program.
12. What is a variable?
13. What two reserved words are used to start and terminate the statements that make up the main body of a Pascal program?
14. The ______________ is used to separate statements in a Pascal program.
15. What is a compound statement?
16. List the keywords to input information into the computer's memory.
17. How many values will be read by the following statement?

    ```
    Read(Length, Width, Height);
    ```

18. Write a Pascal statement to accept four values, named A, B, C, and D, from the keyboard.
19. What keywords are used to output information on a computer's display screen?
20. Write a statement to print the message 33,550,336 IS A PERFECT NUMBER.
21. In a write statement, everything within single quotation marks is ______________ .
22. The statements

    ```
    write(A, B, C);
    write(X, Y, Z);
    ```

will cause the values of A, B, C, X, Y, and Z to be printed on ____________ line(s).

23. The statements

```
writeln(A, B, C);
writeln(D, E, F);
writeln(G, H, I);
```

will cause the values of A, B, C, D, E, F, G, H, and I to be printed on ____________ line(s).

24. Write a statement that will print the integer 693 in a print field of 8 characters.

25. Write a Pascal program that will print your name, address, city, and state on separate lines.

26. What is the value of 50 **div** 10?

27. What is the value of 50 **mod** 10?

28. Write a Pascal statement to add the value 8 to a given variable named Box.

29. What value will be printed whenever the following program is executed?

```
program Exercise29;
var
  a, b, c       :Integer;
begin
  a := 400;
  b := 30;
  c := 128;
  write(a, c, a, b, b, c);
end.
```

30. What value of R will be printed by the following program?

```
program Exercise30;
var
  X, Y, Z, R     :Integer;
begin
  X := 120;
  Y := 200;
  Z := 40;
  R := (X + Y + Z) / 6;
  writeln('Answer = ', R);
end.
```

31. Write a program to input three values of A, B, and C, and output the sums A + C, B + C and A + B.

32. Write a program that will read values for X, Y, and Z and print them, first in the order read and then in reverse order.

33. Write a program to input a length expressed in feet and inches, and output the same length expressed in centimeters. There are 2.54 centimeters to an inch.

34. Write a program to input six real numbers and output their sum.

35. Write a program which will convert a distance quoted in kilometers into miles.

36. Write a program to compute the volume of a sphere, given its radius. The formula for the volume of a sphere is $V = (4\pi r^2) / 3$.

37. A student has taken four examinations. Write a program which reads her four grades (integers) and prints her average grade.

38. Write a program which accepts as input a person's current bank balance and the amount of a withdrawal. The program should print the new balance.

39. Write a program that, given a number as input, prints its square, cube and fourth power.

40. Write a program to compute the diameter, circumference, and the area of a circle of given radius. For a circle of radius r the diameter is $2r$, the circumference is $2\pi r$, and the area is πr^2.

41. Write a program to determine whether Y is between −40 and +40. If Y falls within these limits, print out TRUE; if not, FALSE.

42. Write a program that will display a message if, and only if, an integer typed in from the keyboard is an odd number.

43. Write a program that will display a message to say whether, or not, a number typed in from the keyboard is exactly divisible by four.

44. Write a program which will display the number, in words, of the digit key pressed on the keyboard, i.e., when 2 is pressed, the program will display TWO, etc.

45. Write a program which accepts twenty examination marks and prints a "pass/fail" message depending on whether the average is greater or equal to 60, or less than 60.

46. Write a program to determine which of two numbers entered from the keyboard is the smaller.

47. Write a program to read a set of student test scores, and print out the number of passes and fails. Assume the pass mark is a score of 50 or larger.

48. Write a program which will read the number of a month and, assuming that it is not a leap year, will print the number of days in the month.

49. Write a program that displays one of the following messages if the appropriate key is pressed.

Keyboard Key	Message
A	Message A
B	Message B

C	Message C
D	Message D
E	Message E
F	Message F
G	Message G

50. Simplify this **case** structure as much as you can.

```
case InventoryItemNumber of
  0 : writeln('Bats');
  1 : writeln('Footballs');
  2 : writeln('Baseballs');
  3 : writeln('Gloves');
  4 : writeln('Tennis Rackets');
  5 : writeln('Golf Clubs');
  6 : writeln('Soccer Balls');
  7 : writeln('Basketballs');
  8 : writeln('Boxing Gloves');
end;
```

51. Write a program that counts the number of times the digits 3, 6, and 9 appear in fifty characters worth of input.

52. What is meant by looping?

53. What are the three loop structures for Pascal?

54. Write a **for**-loop that prints your name 40 times.

55. Write a **for**-loop that adds the integers from 1 to 100, inclusive.

56. How many times will the following loop be repeated?

```
for Count := 10 downto 1 do
  writeln(Count);
```

57. How many times will the writeln statement be executed in the nested loop below?

```
for Count1 := 1 to 5 do
  for Count2 := 1 to 6 do
    writeln(Count1 * Count2);
```

58. Write a program that creates a table of Celsius to Fahrenheit temperature conversions from 15 degrees to 35 degrees Celsius. The conversion formula is F = 9/5 * C + 32.

59. Write a program to read in an integer between 1 and 4 inclusive and print out the name of a vehicle with that many wheels.

60. Write a program to find all the integer solutions to the equation

$$A = 4X + 3Y - 2Z + 5$$

for values of X, Y, and Z in the range 0 to 20.

61. Write a program to print the pattern:

```
******
*****
****
***
**
*
```

62. Write a program to read in a list of 10 numbers and print out the smallest number entered.

63. Write a program to read in a list of 8 numbers, all entered on one line, and print them out in reverse order on the next line.

64. Write a program to print out the numbers from 1 to 9 diagonally across the page. In other words, each new value should be five positions further to the right on the next line.

65. Write a program to produce the following output:

```
  1 IS A TRIANGULAR NUMBER
  3 IS A TRIANGULAR NUMBER
  6 IS A TRIANGULAR NUMBER
            •
            •
            •
190 IS A TRIANGULAR NUMBER
```

66. Write a program that prints a list of numbers that are palindromes. (A palindromic number is unchanged if its digits are reversed: 23455432 is a palindrome.)

67. Write a program which reads and adds 10 real numbers and then prints their average.

68. Write a program to print a "5 times table" in the form

```
1 × 5 =  5
2 × 5 = 10
3 × 5 = 15
4 × 5 = 20
      etc.
```

69. Write a program which prints all multiples of 4 from 4 to 100.

70. Write a program which accepts as input two numbers representing the width and height of a rectangle. The program is then to print such a rectangle made up of asterisks, for example:

71. Write a **for**-loop that prints the even numbers between 1 and 25.

72. Take a four-digit number. Add the first two digits to the last two digits. Now, square the sum. Surprise! You've got the original number again. Write a program that will find all four-digit numbers that have this property.

73. Write a program that accepts an integer and determines whether it is a prime number.

74. What are the differences between **while** and **repeat** statements.

75. Write a **while** statement that prints the integers 1 through N.

76. Write a **while** statement to generate a table of integers from −23 through +46 and their squares.

77. Write a **repeat** statement that prints the integers 1 through 100.

78. What is the absolute value of:
    a. 86.9
    b. −86.9
    c. −623

79. What is the greatest integer value for the following numbers?
    a. 9.8
    b. 9.3
    c. 264.05

80. What is the difference between a function and a procedure?

81. Write a program to read in a number and print out its squa root.

82. Write a program that reads an integer N and computes the sum of the squares of the integers from 1 to N.

83. Explain the difference between actual and formal parameters.

84. Explain the difference between value parameters and variable parameters.

85. Explain the difference between local and global variables.

86. What is an array? A subscript? A subscripted variable?

87. Write a program that will put a zero in every element of a one-dimensional array named L. The program should also print the list contents.

88. Thirty integers are stored in a one-dimensional array Score. Exactly one of these numbers is zero. Write a program that finds the zero and replaces it with the number 100.

89. In a two-dimensional array named Table, Table[3, 8] is in the ____________ row and the ____________ column.

[1] 3	[2] 7	[3] 7
[2] 2	2	5
[3] 4	9	6

Answer to crossnumber puzzle on page 131.